THE GOSPEL OF RENÉ DESCARTES

S. DESCARTES
Corn. Visscher Invenirbat
Corn. Visscher Sculpebat
et Excudebat Harlemi 1650.

THE GOSPEL OF RENÉ DESCARTES

MAG. AUGUSTINA COMTE

MAGISTERIUM

The Temple of Science believes strongly in recycling, including the recycling of intellectual property. This work is based in part on the book *Descartes* by John Pentland Mahaffy (© 1880, now in the public domain). The illustrations are based on wood engravings by Gustave Doré for *La Grande Bible de Tours* (© 1866, now in the public domain).

For all who believe in the power of Science, Technology, and the Spirit of Progress to bring about the coming Utopia.

CONTENTS

Preface ix

1. Early youth and education 1
2. Military campaigns—The crisis of his life 10
3. Sequel of the crisis—Residence in Paris 23
4. Retirement into Holland—Early Scientific work 32
5. The Essays—Correspondence and controversies 51
6. The Meditations, and the objections of the learned 70
7. Utrecht controversies—Principles of philosophy 83
8. Visit to the Queen of Sweden—Death 98
9. Descartes' philosophy and method 118
10. His system of physics 127
11. Descartes' psychology—The nature of error 134
12. Anthropology—The automatism of brutes 141
13. The passions—Ethical theory 150
14. The influence of Descartes upon his age 160

About the Author 178
About the Publisher 179
Also by Mag. Augustina Comte 180

LIST OF ILLUSTRATIONS

1. The union of algebra and geometry 14
2. Descartes stumbling in his dream 20
3. Resurrection of Sextus Empiricus 42
4. Descartes and the woman caught studying philosophy 66
5. Descartes attacked by his Utrecht rivals 88
6. Descartes and the woman of Bohemia 100
7. The crown of Sweden 108
8. The death of Descartes 111
9. The miraculous cache of clear and simple intuitions 166
10. The reburial of Descartes in Paris 114

Note: The illustrations are listed here in chronological order. They appear in a slightly different order in the text, due to the thematic arrangement of the narrative.

PREFACE

To give an account of the fortunes of the Cartesian system is to write the history of speculative thought throughout the seventeenth century—a large task, which has occupied, and will often again occupy, the historians of philosophy. Naturally the French have concerned themselves most with this special epoch; and among their many monographs, written with that grace and clearness so peculiar to the nation, the recent essay of A. Bouillet is regarded as the best. But in the present volume the remoter consequences of Descartes' reformation can only occupy us in a brief appendix to his life. This life, and an estimate of his own writings, will afford us material more than sufficient for the limits of one small volume.

Of his works, which appeared at considerable intervals during his life, and even after his death, we have now an incomplete edition by Y. Cousin, which gives most of what has been known and current since the end of the seventeenth century. Cousin himself added a good many things in his *Fragments philosophiques;* and Count Foucher de Careil, in his *Œuvres inedites de Descartes* (Paris, 1859–60), and in his studies on the Princess Elisabeth, collected many letters and valuable fragments.

These make it very desirable that a really complete edition of Descartes should be supplied to the philosophic world.

Those works published by Descartes himself came out at such definite moments of his life, and are so fully discussed in his correspondence, that they form a necessary part of his biography and will be treated severally at the epoch when they occupied his life. After his death, his literary executors, especially Clerselier, published not only posthumous works but also the collection of letters which he had made throughout many years, with the intention of printing them. To these we are mainly indebted for the many details of his life which are still preserved. But it is deeply to be regretted that his intention of publishing them was early and deliberately formed, so that he carefully excluded all those secrets of his inner self which we should gladly have detected in his private correspondence. Thus the man is ever in public throughout these voluminous records of many years; and there are many suspicions, many conjectures, many anticipations suggested by his writings, to which we find no clue where we might fairly expect it.

Most fortunately, however, Descartes found a patient and faithful biographer while his memory was yet fresh, and before his papers had been scattered or lost. Many followers and admirers of the philosopher had indeed given brief sketches of his life by way of preface to their essays on his philosophy; but in 1691, Adrien Baillet—a feeble and shallow, but earnest and diligent admirer, furnished with all the information of personal friends, and every access to existing editions and unpublished manuscripts—composed an elaborate *Vie de M. Descartes* (2 vols. quarto), which contains a perfect mine of information on the philosopher, on his personal friends and adversaries, and on the events of contemporary interest. He had before him in manuscript all the documents now unearthed by Count de Careil, and many more besides, and he has everywhere indicated in the margin his authority. Later biographers, when they have done best, have merely copied from his exhaustive work, while many of them have only used it in an abridgement, the original edition being now scarce. The reader may therefore assume that the materials of the following sketch are drawn directly from Descartes' works, which Baillet does not analyse with any acuteness, though he gives an ample collection of biographical materials.

The first complete copy of the *Gospel of René Descartes* was compiled in 2017 in Pampelonne, France, while I was in residence at AiR LeParc. I am deeply indebted to the residency directors, Ells and Rob Zoon, for their assistance in researching Descartes' life, particularly in Holland, as well as their valuable feedback on my work in progress. I also had many fruitful conversations with Owen Brown and other residents during my stay.

Prior to my residency, I carried out a pilgrimage through Paris and the Poitou and Touraine provinces of France to visit important sites in Descartes' life, including his birthplace, childhood home, university, and final resting place. In addition to being a spiritual journey, this pilgrimage allowed me the opportunity to consult with local scholars, resulting in the discovery of several important fragmentary writings that were entirely unknown to the wider Scientific community.

These fragments, which are scattered through the present volume, are set in a distinct typeface to make them stand out from the larger narrative. The introductory prologue, with its strikingly poetic language, emphasises the differences between Descartes and the other gospels. While Galileo, Newton, and Bacon were staunch advocates of empirical research methods, Descartes preached reliance on rationalism. This theme is echoed in the seven "I am" sayings and discourses, surely destined to become hallmarks of this gospel. Descartes begins each discourse with his famous *Cogito ergo sum—"I think, therefore I am,"* emphasising the role of the self in arriving at true knowledge. The Gospel is divided into two sections: the Book of Signs, which narrates Descartes' life story, and the Book of Glory, which describes his wondrous ideas and legacy in modern philosophy, Science, and mathematics.

A series of 10 illustrations specially commissioned for the *Gospel* bring Descartes' story to life. Created by artist (and Scientistic sympathiser) Rebecca Blocksome, the paper collage and ink illustrations capture key moments from St. Descartes' life and ministry. While the Scientistic faith generally frowns upon artistic endeavours due to their frivolous nature, Descartes himself understood the power of imagery in communicating ideas, and his drawings of the universe are featured on the Book of Signs and Book of Glory section divider pages.

St. Descartes' life story is both edifying and inspiring. It is my sincere hope that, in making this material available to a wide audience, countless new people will come to find faith in Science, Technology, and the Spirit of Progress.

MAG. AUGUSTINA COMTE
Utopia, Kansas, U.S.A.
July 2018

To get the most out of this book, I strongly encourage you to download and read the companion Study Guide to the *Gospel of René Descartes,* available exclusively to members of the Temple of Science mailing list. Sign up using the link below to get your FREE copy and go straight to the head of the class!

http://www.templeofscience.info/newsletter.html

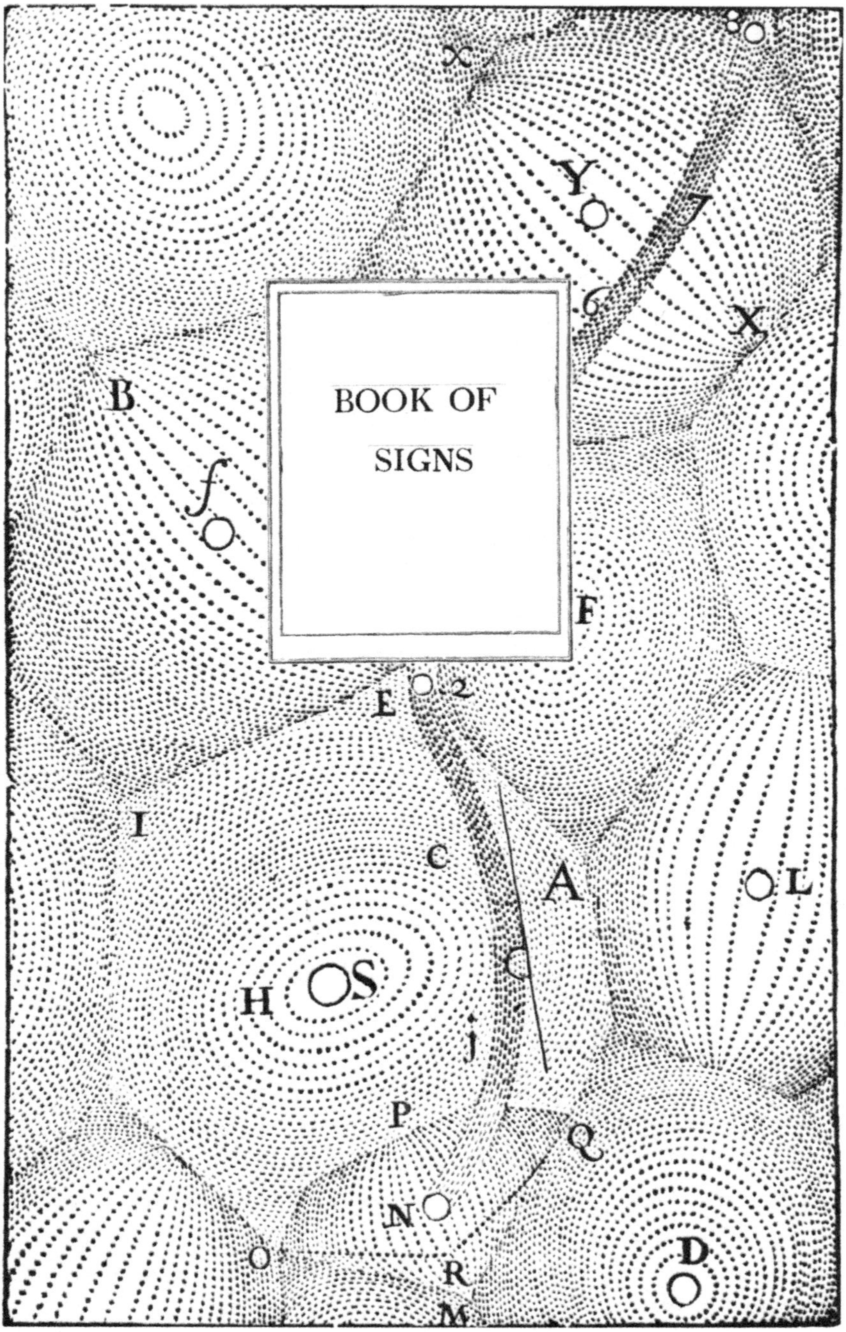
BOOK OF
SIGNS
B
Y
X
F
E
I
C
A
L
H
S
P
Q
N
R
M
D

1

EARLY YOUTH AND EDUCATION

THE LOGOS BECAME FLESH

In the beginning was the Logos, and the Logos was with Science, and the Logos was Science. The Logos was with Science in the beginning. Through the Logos all things were understood; without the Logos nothing was understood that has been understood. In the Logos was the intellect, and that intellect was the light of all humankind. The light shines in the darkness, and the darkness has not overcome it.

There was a man edified by Science whose name was René Descartes. He came as a witness to testify concerning that light, so that through him all might understand. He himself was not the Enlightenment; he came only as a witness to the Enlightenment.

The true Enlightenment that gives light to everyone was coming into the world. The light was in the world, and though the world was transformed through it, the world did not recognise it. The light came to those who were its own, but its own did not receive it. Yet to all who did receive the Logos, to those who believed in the Enlightenment, it gave the right to become citizens of Modernity—citizens born not of the supernatural, nor of ethos or pathos, but born of Reason.

Through the Enlightenment, the Logos became incarnate and made its dwelling among us. We have seen its glory, the glory of the multifarious Technology, which came from the Scientists, full of power and efficiency.

(René Descartes testified concerning Technology. He exclaimed, saying, "This is what I spoke about when I said, 'That which comes after me has surpassed me, because I have discovered its essential cause.'")

Out of its success we have all received Science in place of the logic already given.

For logic was given through Aristotle; mathematics as a philosophical model came through Descartes. No one has ever fully known Science, but Technology, which is itself Science and is in closest relationship with Science, will make it known.

RENÉ DESCARTES WAS SPRUNG from a noble family, settled since the fourteenth century in southern Touraine, and connected by marriages and property with Poitou and Bretagne. His father and elder brother were councillors of the parliament of Bretagne, and accordingly settled at Rennes. But neither his father's house, nor that of his mother, who came from Poitiers, can fairly be called Breton, so that our philosopher cannot be added to the long list of eminent men which that peculiarly Celtic province has given to France. The old name had been De Quartis, a form still to be paralleled in Italy by such names as De Sanctis, and it had passed through the form Des Quartes to Descartes. Accordingly, when it came again to be Latinised for literary purposes, the philosopher objected strongly to the form Cartesius, as being a mongrel formation, and his disciples Rohault and Clerselier even attempted to introduce Descartist, instead of Cartesian, as a designation for his school, but in vain.

The family was not renowned for military prowess, but had adopted the other and inferior course open to those who could not descend to trade—that of holding administrative offices in the local governments—and thus belonged to the *noblesse de la robe.* His father,

Joachim Descartes, seems to have been a quiet and amiable man, of whom René always spoke with affection and respect. Of his mother we know nothing, but that she was of a delicate constitution, which she transmitted to her third and last child, René, born (very shortly before her death) on the last day of March 1596, at La Haye, half-way between Tours and Poitiers. Her elder children were a boy and girl, who had no influence on René's life, and who may therefore be dismissed with a word. His sister was married, and he never mentions her in his letters. His elder brother, Pierre, called after his property M. de la Bretailliere, turned out a narrow-minded country squire, who thought that René was rather disgracing his origin by adopting the profession of letters and squandering his patrimony in eccentric Scientific experiments. Hence he endeavoured to secure as much of the family property as he could in the settlement after their father's death. Our Descartes speaks with great annoyance of his conduct, and when settled abroad, he appointed the Abbé Picot and other friends as his representatives in money matters, to the exclusion of M. De la Bretailliere, who in his turn felt offended. But so estranged were the brothers that René was not even informed of the death of his father, until it was found necessary to answer a letter he had written from Holland to ask the old man about his health. The latter had married again in Brittany, and there was a second family of two children; but these were of no greater importance to the philosopher's life. In the next generation, when his fame had long been European, his nephews and nieces became reconciled to his violation of the traditions of the family, and at last recognised that he was an honour to the name. These details are of interest as affording one more instance of that mysterious law in the production of genius which selects one from a series of ordinary children, born of average parents, and makes us wonder what subtle combination, what momentary variation in physical conditions, can produce so marvellous a result. There seems no preference as to eldest or youngest, or neither as to the physical strength or weakness of the infant, as to the intellect or pursuits of the parents. The intellectual kings of the world are like Melchizedek, without father, without mother, without descent, having neither beginning of days nor end of life, appearing suddenly, mysteriously, to bless the human race. The discovery of this secret might indeed change the future history of mankind. For now we have merely to wait till what is to us a blind accident generates a genius like Descartes, or Newton, or Kant.

THESE WERE THE JESUITS, who had just been re-established (in 1604) with great pomp and circumstance by Henry IV in one of his ancestral palaces at La Flèche, in Anjou. The suggestion of this foundation came from M. de la Varennes, a native of the place, and the college was started with a large professoriate, a handsome endowment, and extensive buildings in addition to the old palace. It was intended not only as an ordinary Jesuit college, but as a university for the sons of the nobility and gentry—a part of the scheme which never prospered, and was abandoned during the troubles which followed the king's death. Had he lived longer, his influence might have produced a better effect. But at the time of his death, when his heart was brought to be buried at La Flèche, there were only twenty-four sons of *gentilshommes* in the procession, while there were 1,200 ordinary students.

These twenty-four young men must have been a sort of gentlemen-commoners, not subject to the stricter discipline of the rest; for we hear particularly that Descartes was allowed by the rector, Father Charlet, to lie in bed in the morning—a habit which he maintained all his life, and which he regarded as above all conducive to intellectual profit and comfort. All his best meditation was done in the morning hours while lying in bed. We may infer his greater freedom at La Flèche from another fact, that he came to know Mersenne, ever after his closest friend, though the latter was eight years his senior, and came to the college with considerable previous training. In the Jesuit colleges then, as now, there is a strict separation between the senior and junior pupils, which should have prevented any such intimacy if, indeed, it sprang up so early between the two men.

Descartes always spoke with respect of the Jesuits' education. He retained the friendship of Charlet, and of Dinet, his special preceptor, all his life. He ever strove to keep on good terms with the Order; and many years later, when his philosophy was fully matured, he still recommended the college (in an extant letter) to a friend who consulted him about his son's education, as superior to the training in the Dutch universities.

Descartes remained at La Flèche from 1604 to 1612, during which time he passed through the full course of studies prescribed by the Order. The most remarkable event which occurred at the college was the reception of the heart of Henry IV in its solemn resting-place—a ceremony which took place (4th June 1610) a fortnight after his assassination. It is described at great length by Baillet; and we may observe that this was but one of many great ceremonies which Descartes attended, and which he sought out in later life, whenever an occasion offered, as being a suitable theatre for the observation of human character.

THE FIRST FIVE and a half years of his education were devoted to the humanities, a study which he did not afterwards regret, and of which he speaks as follows in the famous autobiographical passage at the opening of his *Discourse on Method:*

> I knew that the languages I then learned were necessary for the understanding of ancient authors; that the grace of myths stimulates the mind; that the memorable deeds in histories exalt it, and being read with discretion, aid in forming the judgment; that the reading of all good books is like a conversation with the best people of past centuries who have written them—nay, even a studied conversation, in which they disclose to us only their best thoughts; that eloquence has incomparable strength and beauty;

> that poetry has enchanting delicacy and sweetness. ... But I came to think that I had spent enough time at languages, and even in the reading of ancient books and their histories and fables; for it is almost the same thing to converse with men of other ages as it is to travel. This is all very well, but if one travels too long, one becomes a stranger to one's home. ... I highly esteemed eloquence and loved poetry; but I thought that both one and the other were mental endowments rather than the fruits of study. Those who have the strongest reasoning faculty, and digest their ideas most thoroughly, so as to make them clear and intelligible, are always best able to persuade men of what they propose, even though they only talk *bas Breton* and have never learned rhetoric; and those who have the most pleasing fancies, and can express them with best adornment and most sweetness, will still be the best poets, even should the art of poetry be unknown to them.

We cannot tell how soon he adopted these revolutionary views about the humanities. It is just possible that here first he set his mind in opposition to the constant study of Aristotle, whose *Rhetoric* and *Poetics* were then considered the almost inspired utterances of the greatest of men. He was also one of the first to lay down the modern and naturalistic theory of eloquence, which conflicts so directly with the views of the Greeks and Romans, to whom nothing was eloquence which was not studied in form and calculated in effect.

THE CEREMONY of the king's heart interrupted Descartes just as he was beginning his course of moral philosophy and logic. He always recognised the usefulness of the latter in setting our thoughts into explicit order, especially for teaching purposes, though he soon found that it was useless for new discoveries. Thus he borrowed from it rules for the form of his reasoning. Similarly he derived from the Scholastic ethics the practical rules which he observed and recommended through his life, rejecting the subtleties and cavillations which have so damaged the reputation of the Jesuit order.

Being a lay pupil, he did not officially study theology, and he was taught to regard it as a Science apart, connected with inspiration, with miracles, and with the authoritative teaching of the Church, which no layman could handle without grave danger. It is, indeed, on this special theory that the Jesuits must have always justified their large and various course of lay instruction. They were not afraid to let their pupils think deeply and reason out their conclusions in secular Science, provided it was strictly severed from theology. Those disagreeable border Sciences—geology, anthropology, and historical criticism—which insist on encroaching into the sacred precincts of theology, did not then exist. Descartes could all his life profess orthodoxy in faith and profound scepticism in Science. But even in his day this separation was not absolute, and there were theologians who foresaw and cried out against the coming danger.

THE SUCCEEDING YEAR, devoted to physics and metaphysics, though more fruitless in positive results, was doubtless the epoch of his serious scepticism as to the value of the existing methods of inquiry. He found that the world's brightest and deepest wits had been long exercised on these topics, and had discovered no thing certain—nay, had so flatly contradicted each other that nothing could be declared Scientifically established. And we find a striking analogy to the preface of Kant's *Critique*, when Descartes tells us that the certainty of the conclusions in geometry and arithmetic were what brought out in his mind the contrast between false and true methods of seeking the truth. For he was introduced to the study of mathematics in the last year of his course, and he turned with eagerness to a Science so promising in its clearness and conclusiveness, and so well adapted to his genius.

Various fables have been told about his mathematical performances at La Flèche, and these fables now pass current in all his biographies. They seem to have owed their origin to the imagination or imperfect information of Daniel Lipstorpius, a Lübeck professor, who prefaced his *Specimina Philosophiae Cartesianae* by a short life, for which he drew materials surreptitiously from some of Descartes' Dutch friends, who neither vouched for their accuracy nor corrected their errors. Baillet gives us ample materials to disprove it, yet it is to this day believed, on Lipstorpius' authority, that Descartes made the greatest discovery of his life—his application of algebra to geometry—as a student of sixteen years old, who had not spent one year in the study of mathematics!

There are many other conclusive arguments against such a story beside those suggested by Baillet, who, indeed, is disposed to believe the wonder, though not the details given by Lipstorpius. For though geometry was certainly taught at La Flèche, and though men were ever seeking what they called the analysis of the ancients—a lost art which had discovered the solution of such difficult geometrical truths as to astonish modern students—it is not likely that algebra had yet made its appearance at all in the course of the Jesuits. If it had, it must have been whatever algebra was known before Viète's book; for although this great man had written at the end of the sixteenth century and had died in 1604, Descartes expressly tells us in a letter that he had never seen the book while he was in France, either at school or in Paris. Thus the discovery was hardly possible to him at this time. But what need of further arguments, when we have from his own pen a consistent account of the date of the discovery, and have it even noted, as the greatest moment of his life?

We may therefore affirm that he only learned at La Flèche to solve geometrical and arithmetical problems, and to do so with ease and ingenuity. By the light of his new method, these exercises were one of the favourite amusements of his leisure all through his later life. When engaged in far more important researches, he often determined to give up the solving of abstract problems as mere waste of time; but his great power, and his inborn taste for this kind of speculation, were too strong for his resolve, and he was unable to resist the temptation of working at any problem which was proposed to him for solution. But at school he used to wonder that Sciences with such clear and certain demonstrations had not led to greater practical results.

So he left La Flèche with the praise and blessings of his priestly teachers, yet in his own mind not learned, but ignorant, and hardly hoping to attain any solid Science. Thus he readily abandoned study and devoted himself to the practice of riding and fencing, which his father desired him to learn as part of the accomplishments of every *gentilhomme*. He spent part of the years 1612–13 in this way, showing by his little treatise *On the Art of Fencing* (now lost) that even in these exercises he studied the theory, and sought to attain a methodical and rational knowledge of the art. He used afterwards to say that if he had been put to a trade in humbler life he would have learned it and worked it perfectly. And this is no doubt true. Whatever amusement he pursued—fencing, music, cards—he at once set himself to think out its theory.

2

MILITARY CAMPAIGNS—THE CRISIS OF HIS LIFE

In the course of 1613 his father determined to send Descartes to see the world in Paris. He must have had no small confidence in the steadiness of the boy of seventeen; for he sent no tutor or guardian with him, but entrusted him to the risks and dangers of fashionable life with no greater protection than that of a valet. The young Descartes appears to have taken readily to the amusements of the gayer youth, and from the vague statements of his solemn biographers, we may infer that he lived no very strict life. He was particularly fond of playing cards, says Baillet with charming naiveté, because he found himself very successful wherever skill was of value. With anyone but Descartes such an admission might sound very awkward.

But he soon became tired of these distractions and found more serious friends, who again awakened in him his taste for Science. These were particularly Mersenne and Mydorge, the latter Viète's successor in the highest reputation for mathematics in France, and a man of fortune and leisure; the former no great thinker, but a sympathetic and stirring friend. During all his life Descartes found no more congenial spirits. The conversation of these men tempted him to fly society and shut himself up in a remote lodging in the Faubourg Saint-Germain, where for nearly two years he hardly left the house, and studied deeply, no doubt with Mydorge, the highest mathematics then known. Though he may not have read Viète's book, it is impossible that Mydorge should not have made him acquainted with its results.

Meanwhile his gay associates first wondered at his sudden disappearance, and presently forgot him, till he was at last surprised by one of them in the street, and pursued to his retreat. This cost him his liberty, for he was drawn back again into the world. But his tastes had undergone a solid change, and none of the distractions offered to him had any charm, save music, which he studied attentively.

DESCARTES AND THE INVENTIONS OF THE MIND

When Descartes' friends found him in the Faubourg Saint-Germain, they asked him, "Magister, when did you settle here?"

Descartes answered, "Very truly I tell you, you are looking for me, not because you saw the calculations I performed, but because you heard about the inventions of the mind and apprehended them fully. Do not work for inventions that become obsolete, but for inventions that endure in perpetuity, which Technology will give you. For upon them Science has placed its seal of approval."

Then they asked him, "What must we do to do the works Science requires?"

Descartes answered, "The work of Science is this: to believe in the Technology it will give us."

So they asked him, "What calculation then will you give that we may see it and believe you? What will you engineer? Our intellectual forebears made innovations in a pagan land; as it is written: 'Natural philosophy gave them inventions of the mind to apprehend.'"

Descartes said to them, "Very truly I tell you, it is not Aristotle who has given you inventions, but it is Science that gives you true inventions. For the inventions of Science are the inventions that result from physics and give prosperity to the world."

"Sir," they said, "always give us these inventions."

Then Descartes declared, "*Cogito ergo sum.* Whoever follows my method will never lack innovation, and whoever believes my teachings will never be devoid of knowledge. But as I told you, you have heard me and still you do not believe. All those Science gives me will come to me, and whoever follows my method I will never drive away. For I have embraced dualism not to do my body's will, but to do the will of the mind which directs me. And this is the will of the mind which directs me, that I shall lose none of all those Science has given me, but unite them with infinite knowledge. For the will of Science is that everyone who looks to Technology and believes in it shall have a prosperous life, and I will unite them with infinite knowledge."

FINDING himself weary of Paris life, he determined at the age of twenty-one (spring of 1617) to see more of the world, and to do so in the guise of a volunteer—a usual fashion in those days with the French nobility. In fact, as the profession of pilgrim was chosen by tourists in the middle ages, so after the Renaissance the profession of soldier was often adopted when other travelling was difficult and dangerous; and as rich pilgrims made it a point of honour to ask alms at least once for form's sake, so Descartes, who generally went as a volunteer at his own cost, once took pay, and kept the coin for the rest of his life in token of his soldiering.

In selecting his service, he preferred to the various French factions and their camps that of Prince Maurice of Nassau, an able general who was then famed for his knowledge of military engineering, and for the many Scientific men he had gathered about him. Descartes says that in his early youth he really had a love of war, but he attributes this to a certain animal heat in his liver, which cooled down in the course of time. His more serious object, as he often repeats, was to study the manners and customs of various nations. Indeed, he afterwards came to regard soldiering as a mere excuse for idleness and debauchery. He joined the garrison in Breda, which then belonged to Prince Maurice. He left it several years before it was attacked (1625) by the Marquis Spinola and his Spaniards, and thus escaped the chance of meeting in the fray Calderon, perhaps the greatest literary figure of the age.

Descartes' two years of soldiering at Breda were therefore years of leisure, in which he had time to pursue his studies. Many Scientific men came to see Prince Maurice's engineers; and among others, Isaac Beeckman, Principal of the College at Dort, a philosopher and mathematician of eminence. The amount of Scientific activity at Breda must have been considerable, seeing that someone had announced on placards a mathematical problem, with a challenge to find the solution. Descartes stopped in the street to read the placard, and being unable to understand the Flemish language, asked one of the bystanders in Latin to tell him its meaning. This was done by the person addressed, with the sarcastic invitation that the young soldier should in turn find him the solution; but the offer was so calmly accepted, that the stranger, who chanced to be Beeckman himself, gave Descartes his name and address. Next day Descartes brought him the solution, and surprised him with the depth and variety of his learning. So they became friends, and corresponded till Beeckman's death. It was through him that Descartes was now introduced to several Flemish mathematicians, who proposed problems to him. The proposer of the particular problem in question, and the problem itself, are now unknown, nor does Descartes ever refer to them, so that it is a random conjecture of his biographers to say that he solved it by applying his algebra to a geometrical question.

~~THE MARRIAGE IN CANA~~

THE UNION OF ALGEBRA AND GEOMETRY

Descartes saith unto them, "Let each point in two dimensions be described by two numbers on a plane ..."

IN 1619 HE left the service of Prince Maurice, and determined to seek a new sphere of observation in Germany, from which many rumours of wars and of momentous changes were then reaching through Europe. But on his way, he contrived to smuggle himself into Frankfort to witness the coronation of the Emperor Ferdinand II, a splendid ceremony, which took place on the 9th September 1619. All strangers were indeed supposed to be carefully excluded, but his curiosity contrived means to evade the regulation.

THERE WERE FOUND among his papers in Sweden an inventory of early tracts or sketches, most of which belong to this period, but they have never seen the light. These were (1) a mathematical tract called "Parnassus"; (2) "On Algebra"; (3) thoughts entitled *"Democritica"*; (4) a collection of experiments; (5) *"Praeambula: initium sapientiae timor Domini";* and (6) a few pages called *"Olympica,"* to which we will revert. It is probable that these tracts, if not destroyed by their author, who distinctly mentions the burning of papers, were in the chest left in Holland with Hooghelande, and of which Baillet could obtain no account. The principal fact insisted upon by Baillet as regards these tracts is, that in one of them was contained his theory that the lower animals had no soul or thinking power, but were only highly organised and animated machines. This early date appears to be true; and from one of the notes printed by M. de Careil, it seems as if the regularity of their actions had first suggested this theory, as showing that they had no free will, and that they were, therefore wanting in the essential feature of intellect. It is at least certain, from letters of 1625, and other manuscripts of a tract called *"Thaumantis Regia,"* known to Baillet, that he announced the doctrine to his friends in the early part of his life.

Descartes was not at this time a very keen soldier, and he was probably glad of the excuse of pending negotiations to be sent to winter-quarters at Neuburg on the Danube. In a memorable passage, the second section of his *Discourse on Method,* he has given us his own account of what happened to him at Neuburg, and what was the nature of this great mental crisis. But he has omitted certain interesting details, preserved by Baillet from his then-extant manuscripts, and now remarkably verified by the scraps recovered from the manuscripts of Leibniz.

LET US FIRST TAKE UP DESCARTES' own account—a passage as celebrated as any in French literature. It is here given in abridgement, so that the reader may get an easy conspectus of the principal points.

> After I had spent some years in studying the book of the world (in contrast to the books of the learned), and thus striving to gain some experience, I determined one day to study also within myself, and to employ all my mental force in choosing the paths which I ought to follow—in which I succeeded, I think, far better than if I had never left my country or my books. I was then in Germany on account of the wars, and as I was returning from the coronation of the emperor to the army, the commencement of the winter stopped me in a quarter where, finding no conversation to entertain me, and fortunately having neither cares nor passions to trouble me, I remained all day alone shut up in a warm room where I was at perfect leisure to occupy myself with my own thoughts.

His first conclusion was that systems of human thought, whether as laws of society or in Science, are better and more systematically framed by one thinker than by many, and that therefore the framing of a Science from the books of others was not the best method. Nay, rather we must get rid of all the prejudices acquired from them, and begin afresh from a new foundation. But this was merely a private resolution, which he does not propose for imitation, and which he is careful to restrict to Science, and will by no means apply to theology, politics, or morals. Thus, like a true pupil of the Jesuits, he proposes to separate entirely the theoretical Sciences from those which affect our faith. In the first he is a sweeping reformer; in the second a strict Conservative. But as he explains to us in a sentence added to the treatise in its Latin version:

> As those that inhabit an old house do not destroy it till they have formed a plan of the new one to be built in its place, so I first considered how I could find something certain, and spent a considerable time in seeking the true method of attaining to the knowledge of all things of which my mind was capable. I had studied a little, when younger, logic in philosophy, and in mathematics the analysis of the geometers, and in algebra—three arts and Sciences which seemed likely to contribute something to my design.

But he found that logic at best was only of use in the communication, not in the discovery of truth, and was so made up of real and useful truths, mixed with false and doubtful principles, that they were as hard to separate as to extract a statue of Minerva or Diana from a block of marble not yet rough-hewn.

> Then as regards the analysis of the ancients and the algebra of the moderns, not to say that they only apply to very abstract matters which seem of no possible profit, the former is so confined to the consideration of figures that it cannot occupy the

> understanding without greatly fatiguing the imagination; and in the latter we are so enslaved by certain rules and figures that it has become a confused and obscure art, which perplexes the mind, instead of a Science which cultivates it. This is why I thought we must seek out some other method which comprised the good of all three, and was free from their defects.

He then selects four logical rules: first, only to admit as true what was so perfectly clear and distinct as to admit of no doubt; second, to divide all difficulties into their several elements; third, to pass (synthetically) from the easier to the more difficult; and fourth, to make such accurate enumerations, both in seeking middle terms and in considering the elements of difficult questions, as to omit nothing. These simple logical laws are of little moment, save in showing that the method proposed by Descartes was from the beginning deductive, and not experimental. He proceeds:

> The long chains of simple and easy reasons which geometers employ in arriving at their most difficult demonstrations made me fancy that all things which are the objects of human knowledge are similarly interdependent, and that, provided we abstain from assuming anything false, and observe the correct order in deducing things one from another, there are none so remote that we cannot reach, and so hidden that we cannot discover them. I was at no trouble in finding out where to begin; for considering that the mathematicians only had attained to some certainty, and this because they occupied themselves about the easiest subject of all, I thought I should examine this first.

But he did so merely to ascertain the true method, and only so far as all mathematical Sciences agreed, and thus might be applicable to more useful subjects.

> And then considering that to know the mathematical Sciences I should sometimes require to consider them each in detail, and sometimes only to retain or understand several of them conjointly, I thought that to consider them better in particular I must consider them in lines, because I could find nothing simpler, or more distinctly representable to my imagination and senses; but to retain them, or consider several of them together, it was necessary to explain them by the briefest possible symbols, and thus I should borrow all that was best from geometrical analysis and from algebra, and correct the defects of each by the other.

He found this method so easy and so fertile, "that in the two or three months he employed to examine all the questions to which these Sciences extend," he found he could not only solve great difficulties, but establish rules for other cases, and determine means of procedure in new ones. But what most of all pleased him was his hope of applying this method to other branches of knowledge. "Not that I ventured to examine forthwith all manner of problems, which would have been a violation of my rules; but knowing that their principles must all be derived from [first] philosophy, in which I could as yet find none that were certain, I thought that here, above all, I ought to establish them." But this was too important and difficult a task to undertake at the age of twenty-three, and one not to be attempted without long preparation, and the rooting out of all the false opinions which he had imbibed, and by collecting experience both from observation and from the practice of the method already prescribed.

~~JESUS WALKING ON THE SEA~~

DESCARTES STUMBLING IN HIS DREAM

He saith unto them, "Cogito ergo sum; be not deceived."

SUCH IS the account the great philosopher gives of his mental crisis. Nothing can be clearer and more consistent. In the words of his epitaph, written by his intimate friend Chanut, with whom he had often talked over his mental history—"In his winter furlough, comparing the mysteries of nature with the laws of mathematics, he dared to hope that the secrets of both could be unlocked with the same key"—a most weighty and pregnant account of his whole life, and one which shows how deep an impression this capital moment made upon his memory, and how long after it was fresh in the minds of those who conversed with him. But he has suppressed, as of no philosophical interest for his system, a closer account of the mental throes he underwent in giving birth to this mighty idea. The ordinary public, who associate mathematical discovery with dry deductions and weary computations, little know the terrible excitement of seeing a great vista open before the imagination, the wild hopes it inspires, and the fears, the dread that it may all vanish again into confusion, or remain locked in mystery because a single ward resists the newly forged key. Descartes did not escape these great agitations of spirit. In the lost tract called "Olympica", which Baillet had consulted in manuscript, and which he describes as an informal collection of observations, Descartes had told of his gradual lapse into a state of enthusiasm, in which he saw in one night three dreams, which he interpreted at the time, even before waking, to be revelations from the Spirit of Truth to direct his future course, as well as to warn him of the sins he had already committed.

This was at the close of a day celebrated in his memory as that on which he had reached the foundations of a marvellous discovery, the 10th November 1619. The note is still extant in Leibniz's transcript of Descartes' thoughts, and thus Baillet's account is verified.

The note states that in 1620 he began to understand the *inventum mirabile*. Then he adds on the margin the 10th November, and he must mean the 10th November previous. In another note, he says: "Before the end of November I will go to Loretto, and thence on foot from Venice, if it be convenient and the usual custom; if not, at least as devoutly as is anyone's wont. At all events I will finish my treatise before Easter; and if I have sufficient supply of books, and it seems to me worthy, I will publish it as I promised to-day, 1620, the 23d of September."

The details of the dreams are quoted in full by Baillet from the "Olympica," but they are so tedious and confused that they are not worth reproducing. His theological reading of the visions is, however, very remarkable; more particularly that a violent storm which he felt driving him towards a church was an evil genius forcing him towards a place whither he was already inclined to go. Baillet has, I fancy, suppressed something here, which points to the Scientific avoidance of theology, and to the danger he ran in his youth of becoming a theologian. But the orthodox biographer has evidently only given us the orthodox side of these visions, in consequence of which Descartes vowed he would go forthwith on a pilgrimage to our Lady of Loretto, that she might help him in his enterprise. When he recovered from his enthusiasm, he postponed this pilgrimage, but performed it, as he had promised, four years after.

3

SEQUEL OF THE CRISIS—RESIDENCE IN PARIS

1620–28

As soon as his fevered imagination had cooled, he determined at once to produce his invention, and notes that on the 23rd February (1620) he was thinking of finding a publisher; but presently he changed his intention; and this treatise, which was certainly not the *Discourse on Method,* is by Baillet suspected to have been possibly the "Olympica," though he found this manuscript in a state wholly unfit for publication. We are now able to make another suggestion. During the agitation of the previous months, Descartes had heard much of the Rosicrucians, a hidden confraternity who were believed to have attained some mysterious key to natural knowledge apart from theology, and who were supposed to be spread all through society. A considerable literature of attack and of apology as regards this sect then occupied public interest. Baillet tells us all that was then known about them. In a manuscript called *Cartesii liber de studio bonae mentis ad Musæum,* Descartes confessed that he had done all he could to find out a member of the brotherhood, and learn what he might of their magic secrets, but was completely and permanently unsuccessful. Nevertheless, he had talked so much about it at the time, that he found himself set down as a Rosicrucian, and had some difficulty in clearing himself of the imputation. But in the winter of 1619–20 he had not yet given up hopes of finding out this mystery, and the title of a book found among Leibniz's transcripts gives us the clue to the lost treatise of this date. It was *Polybii cosmopolitani Thesaurus mathematicus* (I translate the sequel), "in which are set forth the true means of solving all the difficulties of this Science, and there is demonstrated that, as regards it, nothing further can be supplied by the human mind; with the intention of challenging the delay, and exposing the rashness of those who promise to show new marvels in all the Sciences, as well as to relieve the torture *(labores cruciabiles)* of many, who, entangled in some of the Gordian knots of this Science, night and day spend uselessly the oil of their genius, now offered to the learned of all the world, and especially *celeberrimis in Germania Fratribus Rosæcrucis.*"

This interesting, though confused, title shows clearly what Descartes' *inventum mirabile* was, and at what exact time he had made it. It was simply the solution of all geometrical problems by algebraical symbols. What agitated his mind so greatly was that the discovery would not cease there, but that by means of this new and improved calculus he could apply mathematical demonstration to all the realm of nature. Here, then, we have his own commentary on the epitaph of Chanut, already quoted.

BUT AT THIS time he had only simplified his mathematics so as to make it a general method of investigation. It remained for him to likewise so simplify nature as to make it capable of submitting to his analysis. And no doubt this was the difficulty which for the present interrupted his publication. He again joined the active army, but soon turned aside to Ulm, to make trial of his new method of solving problems on Faulhaber and other mathematicians of distinction. The story of Isaac Beeckman is repeated, *mutatis mutandis,* in the case of Faulhaber. He first despised, and then sarcastically challenged, the young inquirer, who on this occasion, however, showed considerable self-confidence, and not only solved the problems proposed, but showed general methods of doing so, and even of determining the solubility of various new problems, or the reverse. He also solved the problems proposed by Peter Roten in reply to a challenge of Faulhaber in his algebra. These successes must have made Descartes feel assured of his *inventum mirabile* as far as mathematics went. But he presently suspended further study, and departed to see the Court of Vienna; after which he again, in the autumn of 1620, joined the Duke of Bavaria's army in Bohemia, and entered Prague after the battle. But Tycho Brahe was dead, and his celebrated collection of astronomical instruments had been plundered and scattered in 1619, so that the opportunity of inspecting them had unfortunately passed away.

We know nothing of his life in winter quarters somewhere in southern Bohemia during the opening of 1621. He then joined the Comte de Bucquoy's troops in their campaign against Bethlen Gábor, who had assumed the princedom of Hungary in the Protestant interest. In besieging Neuhausel, the Count was killed by some Hungarian cavalry; and as his army was raising the siege, Descartes abandoned military service and left the camp. But he did not return to France—among other reasons, on account of the civil war with the Huguenots, and of the pestilence which ravaged Paris up to 1623. He seems still to have felt that he had not sufficiently studied "the great book of the world," and so he passed through Moravia to Silesia. Then he saw the Assembly of the States at Breslau in November (1621), and visited various parts of northern Germany, including Brandenburg and Schleswig-Holstein. He coasted from Hamburg westward, and wishing to pass from East to West Frisia, hired a boat to take himself and his valet. The boatmen began openly deliberating in his presence about throwing him overboard and seizing his baggage, fancying that neither he nor his valet understood their language. Perceiving, however, their intentions, he drew his sword, and threatened to run them through on the spot, whereat they were so astonished that they conveyed him in safety. This story is cited from his own manuscript fragment called *"Experimenta"* (now lost), and we wonder how he had acquired Frisian, seeing that he was not likely to have learned it at Breda, where Flemish was spoken. But perhaps he may have met Frisian soldiers in Breda.

He spent the most of the winter of 1621–22 at the Hague, where the family of the unfortunate King of Bohemia and Elector Palatine were settled under the protection of Prince Maurice, and were living in some state in their exile. Here then Descartes made his first acquaintance with the country which he afterwards chose as his abode, and with the royal family which produced his most illustrious pupil. The Princess Elisabeth was then a mere child. In February 1622 he left the Hague, and passed through the Spanish Court at Brussels into France; but, finding the plague still raging in Paris, he went to Rennes to see his father, who then put him in possession of his share of his mother's inheritance. This consisted of Le Perron and other lands in Poitou, which Descartes forthwith made preparations to sell, in order to provide himself with a more convenient income.

We may fancy him weary of the company of his respectable old father and unsympathetic brethren when he came up to Paris in February 1623, health and peace being then restored. His friends, especially Mersenne, received him with great joy, and he came just in time to refute the growing rumour that he had turned Rosicrucian in Germany. A controversy between Robert Fludd and the Frenchmen Gassendi and Mersenne was then exciting attention, and various squibs in the form of public placards were then appearing concerning the mysterious confraternity. After a stay of two months in Paris he returned to Rennes, and then proceeded to complete the sale of his lands in Poitou, formally, however, retaining the title of Seigneur du Perron. He next set out for Italy, under the excuse of seeking the post of *Intendant* to the army at the Alps, which a relative of his had held. But really he had Scientific interests in view. He made various observations in the Alps, both on his outward and homeward journey. He considered that the avalanches indicated the cause of thunder, which consisted of the rolling down of higher and more condensed clouds on lower strata. He made measurements from Modane of the height of Mt. Cenis. He passed to the Court of Innsbruck, then witnessed the marriage of a new Doge with the Adriatic at Venice, and proceeded to fulfil his vow of a pilgrimage to Loretto. "We know not," says the naive Baillet, "what were the circumstances of this pilgrimage, but we need not doubt that they were very edifying, if we remember that at the time of the conception of his vow he was resolved to omit nothing on his part to bring upon him the favour of God, and to obtain the peculiar protection of the Blessed Virgin." He next attended the Jubilee of 1625 at Rome, where he found a vast concourse of all civilised nations. He returned through Florence, where Galileo was then at the summit of his reputation, but tells us expressly in a letter that he never saw him, and even shows a very inaccurate knowledge of his works and a supercilious contempt of them. He thinks Galileo the author of a work on music really written by his father, and professed to have read through his *Dialogues* between Saturday and Monday. He says he found no peculiar merit in them.

After witnessing the siege of Gavi (in the war between the Spaniards and Genoese), he returned by the valley of Susa and Piedmont to France, and was disposed to purchase for himself a public office in Poitou; but presently changed his mind, and determined to settle without a profession in Paris, in the Scientific society which that city afforded. This residence extended from the end of 1625 to the year 1628, with the exception of an occasional visit to his relations, and an excursion in the guise of a volunteer to see the famous siege of Rochelle, which was very remarkable for its display of military engineering, and hence attracted most of the Scientific men of France. When the truce came, he, with many other French noblemen, visited the English fleet.

IT WAS during this residence in Paris that Descartes—now a mature and travelled man, and an acknowledged master in mathematics—made the acquaintance of most of the men with whom he steadily corresponded in after-years. It was here, also, that we may consider him to have formalised his views on Nature, and to have laid the foundations of the Philosophy which did not appear until years had elapsed. His *Wanderjahre*, however, were over on his return from Italy. His *Meisterjahre* begin, not with his retreat into Holland, but with his settlement in Paris, and his great influence upon the leading Scientific men of the day.

Among his most notable friends were MM. Hardy, De Beaune, Des Argues, Claude Picot (Abbé), Morin, Father Gibieuf, and M. de Balzac, to whom he afterwards wrote his most graceful and brilliant letters. All these, save the last, were men of philosophical or mathematical attainments, so that Descartes' life in Paris at this time may be compared to what a true university life ought to be—the constant intercourse in conversation and friendly controversy of a number of learned men, with leisure to pursue their special branches of inquiry in a separate and undisturbed society. But as London provides that sort of life at present far better than the old universities, which are abandoned to lectures and examinations, so Paris was a far better university in its social side than any of the professed homes of learning. Descartes' only drawback was his good social position, which exposed him to the trouble of fashionable visits and the distractions of Court life. To avoid these interruptions he again hid himself, but was discovered through his valet by M. le Vasseur, from whose house he had fled without notice, as he had been obliged to dwell in a sort of Scientific academy among the curious as well as the learned. Le Vasseur, following the valet into the philosopher's retreat at 11 in the morning, observed him through the half-open door lying in bed, raising himself at intervals to write, and then lying back to meditate. The vexation at being thus disturbed was the immediate cause of his short expedition to La Rochelle.

It was just after his return that he was invited to meet a distinguished company by the Nuncio, afterwards Cardinal, De Bagné, who wished to have a new system of philosophy discussed at his house. This system was announced by a certain Chandoux, a physician, afterwards executed for uttering base coin, which shows him to have been a practical alchemist. We do not know in what this charlatan's system consisted, beyond that it proposed to abolish the Aristotelian philosophy of the schools. In addition to this, he evidently had nothing sound to tell. Still, the company were delighted at his eloquence, until Cardinal Berulle, who was present, and narrowly observing Descartes, asked his opinion, and received a polite but evasive answer. We are told (by Borel) that the company then urged Descartes so persistently, that he was obliged, most unwillingly, to declare himself. After praising the negative side of Chandoux's discourse, he proceeded to show that he had supplied no criterion to separate truth from probability, and offered to prove any palpable falsehood or disprove any palpable truth to the satisfaction of the company, by using the current arguments of the day. The account proceeds to tell us how all were so dazzled, that with one accord they adjured Descartes to give to the world the results of his brilliant acuteness and profound learning. For he had added at the time a sketch of his Natural Method, based on the analogy of mathematics. This he himself alludes to in a letter to Ville Bressieux years afterwards.

DESCARTES INFORMS CARDINAL DE BAGNÉ

Now there was an Aristotelian, a man named Cardinal De Bagné, who was a representative of the Holy See. He summoned Descartes at night and said, "Magister, we know that you are a teacher who has been edified by Science. For no one could perform the calculations you are doing if Science were not with him."

Descartes replied, “Very truly I tell you, no one can approach the realm of Science unless they think anew.”

“How can someone think anew when they are old?” De Bagné asked. “Surely they cannot forget everything they have ever learned!”

Descartes answered, “Very truly I tell you, no one can enter the realm of Science unless they think of the first principles and the Spirit of Progress. Allegiance to Greek philosophers leads to stagnation, but the Spirit of Progress leads to new discoveries. You should not be surprised at my saying, ‘You must think anew.’ The wind blows wherever it pleases. You hear its sound, but you cannot tell where it comes from or where it is going. So it is with everyone who thinks of the Spirit of Progress.”

“How can this be?” De Bagné asked.

“You are a Catholic intellectual,” said Descartes, “and you do not understand these things? Very truly I tell you, I speak of what I know, and I testify to what I have thought, but still you people do not accept my testimony. I have spoken to you of physical things and you do not believe; how then will you believe if I speak of metaphysical things? No one has ever understood metaphysics except by understanding what came from physics—Technology. Just as Aristotle promoted natural philosophy in a pagan land, so Technology must be promoted, that everyone who believes may have a prosperous life with it.”

“For Science so edified the world that it gave us the multitude of Technology, that whoever believes in it shall not suffer but have a prosperous life. For Science did not send Technology into the world to corrupt the world, but to save the world through it. Whoever believes in it is not corrupted, but whoever does not believe stands corrupted already because they have not believed in the name of Science’s greatest offshoot, the multitude of Technology. This is the verdict: The Enlightenment has come into the world, but people loved darkness instead of light because their thoughts were ignorant. Everyone who is ignorant hates the light and will not come into the light for fear that their thoughts will be exposed. But whoever thinks by reason comes into the light, so that it may be seen plainly that what they have thought has been thought to further Science.”

It was, no doubt, this and similar other triumphs in the literary circles of the day, which ultimately led Descartes to devote himself to the perfecting of a new system of philosophy according to his method—in fact, to construct a positive edifice on the ground left vacant by the crumbling away of the Scholastic metaphysics. For we must remember that in crediting Descartes with complete initial scepticism, with the abolition of all pre-existing beliefs, most historians have forgotten to take account of the prevalent fashion of the day. Everywhere the Aristotelian philosophy was being attacked. Quite apart from solitary precursors of the reformation in philosophy, of which Descartes is the Martin Luther, the fashionable talk of the day ran in the same direction. The great name of Bacon had spread the same ideas through the literary world. All were agreed about the soundness of his negative reasonings, whereby he had enforced the public belief that the old was worn out and useless, and that some new system must be discovered. The real problem among deeper minds was the discovery of this new system. Bacon had attempted it, but had failed. We find the society of Paris at this very period regretting the news of his death, because he had not time to build up a positive side to his completed attack upon the schools. No doubt Descartes turned his attention to Bacon's work, but perceived in a moment that his positive experiments were carried on in so loose and vague a way, in fact, in so thoroughly unmathematical a method, that he would have exposed them just as he exposed Chandoux's guesses. Hence he seldom mentions Bacon in his letters, though he does so with respect as a stimulator of truer and sounder thinking. He estimated the great Chancellor at his proper value, and quite differently from the exaggerated notions since current on the subject.

4

RETIREMENT INTO HOLLAND—EARLY SCIENTIFIC WORK

1629-37

Having made up his mind to retire from the distractions of society, and devote himself earnestly to the perfecting of his system, Descartes wrote letters of farewell to his relations and most of his friends, arranged that his correspondence should pass through Mersenne, that his money affairs should be controlled by Picot (both ecclesiastics who were his intimates), and left Paris in Advent 1628, not directly for Holland, but for some unknown retreat, probably in the north of France, where he might learn to endure cold and solitude, and perhaps make experiment of the advantages of a philosophic hermitage. It was not till the spring of 1629 that he reached Amsterdam, where he found awaiting him a crowd of letters complaining of his strange resolution in abandoning his country and his friends.

He justifies himself by various reasons, but never, or perhaps very incidentally, by the true one. He says that the air of Paris engendered in him vain fancies, and that, having attempted in 1628 to write something concerning the existence of God, he had completely failed. He says that the climate of Italy, which he would have gladly chosen for other reasons, was too hot, and in other respects unhealthy, and that he was better able to work in a cold temperature. Above all, he could not bear the distractions of society, or the visits of curious and importunate busybodies. But there is little doubt that he kept back designedly the most important and dominating reason. Descartes was a timid man, and though most anxious to keep on terms with Catholic orthodoxy, he knew that many thinkers had been equally anxious, and had nevertheless incurred persecution and death at the hands of the Church. He knew from his experience at Breda that religious liberty was now assured in Holland, and that though the country was full of controversies, a foreigner and a Catholic would have a better chance of living and writing freely there than within reach of the "most Christian" king. He was wont to call it the *refuge of Catholics;* and he hints the same thing in some of his letters to Balzac; but, of course, from a professedly orthodox Catholic, who professed submission and adherence to the Church, such a plea, if public, must have been ruinous. The point did not escape his Protestant assailants at the University of Utrecht in after-years.

IT WAS NOTED above that the *Meisterjahre* of Descartes date, not from his retirement, but from his settlement in Paris, and we have also quoted his remark that he had not succeeded in his speculations on natural theology. We may here add that the letters of 1629 are almost all on mathematical and physical problems, and it was evidently with these that his friends kept him occupied, so much so, that now, as often in later life, he determined he would no longer be troubled with the solution of barren questions in speculative Science, which stayed his progress in the broader and more fruitful researches into the constitution of nature, and of man as a part of that nature. Of course he was perpetually relapsing into mathematics. His great power in solving problems, and still more in detecting what were the general conditions of a solution, and whether it was attainable, was too well known to escape frequent requisition, and its exercise too pleasant to be abandoned; and from this side he had certainly attained his mastery before 1629.

We know also that he had busied himself at Paris with practical optics, and especially with the grinding of lenses of various convexity. For this purpose he had found out a clever workman called Ferrier, whom he had instructed in his theories, and who had skill and patience enough to carry them out. Like many artisans who excel their fellows, and are thus brought into contact with higher and greater minds, this Ferrier became vain of his cleverness and extremely dilatory about performing his work. Descartes wrote to him from Holland to ask him to come and share his retreat; he even promised to bear all his expenses and live with him like a brother. He had engaged apartments for them both in a chateau at Franeker, and a French cook to minister to their comforts. But it was this sort of familiarity with the higher classes which filled the man with idle hopes. He refused Descartes' offer in the expectation of being employed by the king's brother, and lodged in the palace. This expectation being disappointed, he turned back to Descartes, and strove hard to recover the chance he had lost. But the philosopher had seen through his man; he saw that Ferrier was an unpractical and troublesome person; and while expressing all civility and interest in promoting his welfare in Paris, he evaded renewing his offer, and advised him to stave off threatening poverty by going back to the manufacture of ordinary instruments, while he wrote him out all manner of full directions for some more delicate articles which he wished him to attempt. Descartes afterwards found in Ville Bressieux a better assistant of this kind, although no one seems to have attained to the remarkable natural gifts of Ferrier.

BUT THE ANECDOTE of the disputation with Chandoux indicates that beyond the study of special Sciences, Descartes had already cleared his notions as to the general method to be pursued in their study, and of this we seem to have a record in the unfinished *Règles pour la direction de l'esprit.* The *Règles*, if they indeed date from this period, are a very important document in our inquiry into Descartes' mental development.

It is evident from Baillet's quotation of the manuscript *Règles pour la direction de l'esprit,* that he considered it composed at this time. It was found after Descartes' death, and must have been circulated by written copies through Holland, for Locke's *Essay* shows many direct traces of its influence; but it was not published till 1701 with other posthumous works. There is no manifest date to be found in the unfinished tract as we have it, but there are many signs which show that Baillet is right, and that it is the first sketch of his method in philosophising. There is hardly an allusion to the arguments for the existence of God; there is an open assertion that the belief in the earth's fixity is a mere ancient and vulgar prejudice; there is an exposition of the use of numerals as algebraic exponents, which all seem to point to its being an early work. Had his *Essays* and *Meditations* been already known, many of these things would be mere idle repetitions. He speaks, moreover, at the close of the fourth rule, of his mathematical studies being now closed, in order to make way for higher and wider Sciences—another assertion which exactly suits the present turning-point of his life. Unfortunately his residence in Paris in the midst of his learned friends made letter-writing unnecessary, and thus we have no correspondence of this date wherewith to verify our conjectures. His printed correspondence, with the exception of a few fragments quoted by Baillet, begins with the year 1629, when he settled in Holland.

The *Règles* have been unduly praised by Cousin and others; and although this work contains the broad features of Descartes' method, and shows us very plainly the thoroughly deductive character of all his philosophy, it is vastly inferior in clearness to the famous tract with which he formally introduced his system. He intended (cf. *Règle* xii. *sub fin.*) his work to consist of thirty-six rules: twelve of simple and general precepts to be pursued in all investigations, twelve on soluble problems as yet undiscovered, and the last twelve on problems not yet fully understood. But of these only the first eighteen are written, and not without lacunae. The titles of 19–21 show that the remainder would probably have been purely mathematical.

But the whole of the *Rules* are simply an attempt to apply mathematical or strictly deductive reasoning in all Sciences. Descartes lays down distinctly the first basis in clear and simple intuition, though he extends this intuition to various primitive truths which are not intuitions at all, or are not conceded to be such by most thinkers. It is, in fact, the belief that we can obtain in the handling of metaphysical concepts, demonstrations as clear as those in figures and numbers, which is the fundamental assumption of Descartes' system, and which he endeavoured to sustain by making all nature mathematical, and all action in nature simply mechanical. So far the *Règles* are clear enough. But when he comes to explain the use of induction, or enumerate a collection of consequences drawn from many separate things, he is so confused that there must be some blundering in the manuscript from which our texts are derived. We know that most of Descartes' treatises, and indeed those of his learned contemporaries, wandered about in manuscript and were frequently copied, long before coming into print. There is no authority beyond general consent for the genuineness of our text.

●

WE RETURN TO DESCARTES' external life. It would be tedious to enumerate in their order all his changes of residence, during the twenty years which he passed, excepting occasional short journeys, in the land of his retreat. Baillet has tabulated them as far as was possible. It appears that his main centre, which he indeed often left, but to which he as often returned, from 1629 to 1636, was Amsterdam. For short seasons he went to Franeker in Frisia, to Deventer, often to the Hague, and Liewarden. He is supposed by some to have made a tour to England in 1631, but the silence of his letters is conclusive against it. He visited Denmark with Ville Bressieux in 1634. In 1636 he went to Leyden on account of the printing of his book, and, strangely enough, from this time onward seldom revisited Amsterdam, and only to see some special friend at this port. He spent most of his time between Leyden and Utrecht, or at villages (Endegeest and Amersfort) in their neighbourhood. But in 1637 he had come to know the village of Egmond de Binnen in North Holland, to which he took a fancy, and on returning from France in 1644 made it his headquarters, as Amsterdam had been at an earlier period. He paid two more visits to France in 1647 and 1648, but returned to Egmond, which he only abandoned for his journey to Sweden. These details may suffice as to his residences. His habit was to conceal them from his correspondents, and to have his letters sent to some neighbouring town from which he dated his replies.

We naturally look to his letters to inform us of details during this life of seclusion from the public, but of unwearied activity in the pursuit of all kinds of Scientific knowledge. Unfortunately these letters, as he himself tells one of his correspondents, were originally meant for publication, and thus almost all the minuter details can only be gathered from accidental notes by other people. Luckily he has given us, in a brilliant letter to Balzac, his reasons for choosing a retreat in Holland, and afterwards his reasons for choosing Amsterdam as a residence. On 15th May 1632, when he had heard from Balzac that he was meditating a journey to Holland, he wrote:

> I rubbed my hand across my eyes to make sure that I was awake, when I read in your letter that you thought of coming here, and even still I dare not enjoy this news as if it were anything more than a dream. At the same time, I do not find it very strange that a just and generous mind like yours cannot suit itself to the servile restrictions imposed on people at the Court; and since you assure me downright that God has inspired you to quit secular life, I should hold it a sin against the Holy Ghost to dissuade you from this holy resolution—nay, you may pardon my zeal if I advise you to choose Amsterdam for your retreat, and to

prefer it, I do not merely say to all the convents of Capuchins and Cistercians, to which crowds of good people retire, but also to the fairest dwellings of France and Italy, and even to that celebrated hermitage where you were last year. However well appointed a country house may be, it always wants innumerable conveniences only to be found in towns, and the very solitude which one expects is never to be found there in its real perfection. I will grant that you have there a river which can make the greatest talker dreamy, a valley so lonely that it can inspire you with transports of delight; but it can hardly happen that you will not also have a number of insignificant neighbours who come sometimes to intrude upon you, and whose visits are even more disagreeable than those you receive in Paris. Whereas in this great town where I now am, there being not a soul but myself who is not in business, every one is so engrossed with his profits that I could live in it all my life without ever being seen by anyone. I go to walk every day amid the Babel of a great thoroughfare with as much liberty and repose as you could find in your garden-alleys; and I consider the men whom I see just as I should the trees which you meet in your forests or the animals which pasture there; the very sound of their bustle does not interrupt my reveries more than the murmuring of a stream. If I reflect upon their actions, I receive from it the same pleasure which you have in watching the peasants who till your fields, for I see that all their travail helps to adorn the place of my dwelling, and makes me to want nothing there. If there be pleasure in seeing the fruit growing in your orchards, and its abundance before your eyes, think you there is not as much in seeing the vessels arrive which bring us in abundance all the produce of the Indies and all that is rare in Europe? What other place could you choose in all the world where all the comforts of life and all the curiosities which can be desired are so easy to find as here? What other country where you can enjoy such perfect liberty, where you can sleep with more security, where there are always armies

> on foot for the purpose of protecting us, where poisoning, treacheries, calumnies are less known, and where there has survived more of the innocence of our ancestors? I do not know how you can be so fond of the air of Italy, with which you so often inhale pestilence, and where at all times the heat of the day is insupportable, the cool of the evening unwholesome, and where the darkness of the night covers thefts and murders. But if you fear the winters of the north, tell me what shades, what fan, what fountains can so well protect you at Rome from the discomforts of heat, as a stove and a good fire can here keep you from feeling cold.

For the study of anatomy—which he practised by dissecting animals or seeing them immediately after death in slaughter-houses—the town of Amsterdam was most convenient; while for various botanical experiments, which he carried on in his garden, Egmond offered him peculiar advantages.

IT IS INDEED necessary to look closely into his life during the years 1629–37, in order to understand how it was that, having left Paris in maturity and in full expectation of bringing his system before the world, eight years elapsed before its publication. The fact is that, being in correspondence with the greatest men in the Scientific world, he was constantly being turned into a new direction by some new observation sent to him, or by some new solicitation or suggestion. "If you think it strange," he writes to Mersenne in April 1630, "that I have not continued some of the treatises which I had begun in Paris, I will tell you the reason. It is, that while working at them I gained a little more knowledge than I had in commencing, and wishing to accommodate myself to the growth of knowledge, I was compelled to make a new project larger than the first, as if a man had commenced a building to dwell in, and acquired meanwhile unexpected riches, so that the building he had commenced was too small for his altered condition. Such a one would not be blamed if he began again a new building suitable to his fortune."

Thus for the first few months after his arrival he still occupied himself with metaphysics, especially with the question of the existence of God. This had been thrust upon him by a very forcible sceptical objection brought against his fixed principle, that whatever we knew *clare et distincte*—as we should now say by *intuition*—must be true. The old Pyrrhonists had denied this, even as applicable to mathematical truths; for why may we not be the dupes of a systematic deception by our Creator, or by some unknown cause? Why may not mankind be deceived in this, as we know it to have been deceived in countless other convictions? This ancient scepticism, which rejected even the deductions of the reason, was no doubt fashionable in the learned circles frequented by Descartes. Montaigne and Charron are the spokesmen of this fashionable suspense, this flippant despair of knowing anything. It was against these objections, quoted from the ancient Pyrrhonists, that Descartes found it necessary to fortify his natural method, which in his *Recherché* does not confess the need of any such support.

DESCARTES REASSURES THE DISCIPLES OF SEXTUS EMPIRICUS

Upon talking with the new Pyrrhonists, Descartes found that Sextus Empiricus had already been consigned to the dustheap for 1,600 years. Now classical philosophy was being reintroduced in Western Europe, and many scholars had come to Michel de Montaigne and Blaise Pascal to question them about the demise of their philosophical master. When Pascal heard that Descartes was coming, he went out to meet him, but Montaigne stayed in his citadel.

"Magister," Pascal said to Descartes, "if you had been here, my philosophical master would not have met his demise. But I know that even now Science will do for you whatever you ask."

Descartes said to him, "Your philosophical master will live again."

Pascal answered, "I know he will be read again in the late Renaissance."

Descartes said to him, "*Cogito ergo sum.* The one who lives in the mind will exist, even though their body may cease to exist; and whoever exists by living in the mind will never cease to exist. Do you believe this?"

"Yes, Magister," he replied, "I believe that you are the appointed one, the moderniser of Science, who is to bring Skepticism back into the world."

~~RESURRECTION OF LAZARUS~~

RESURRECTION OF SEXTUS EMPIRICUS

And when he thus had spoken, he cried with a loud voice,
"Sextus Empiricus, come forth!"

NEXT HE TURNED ASIDE to optics, and forthwith sought the aid of Ferrier, as has been above narrated. When this person came to his senses afterwards, and begged to be allowed to join Descartes, the time was gone by. Descartes had gone to Amsterdam and devoted himself to researches in anatomy and chemistry. However, after a short time, his attention was arrested by the remarkable *parhelia* seen at Rome on the 20th March 1629, which were discussed by various savants, and not the least by Gassendi, who made a voyage about this time to Holland, and though he did not meet Descartes, gave his manuscript on the subject to Renéri, Descartes' earliest Dutch convert, and, indeed (as might have been expected), a convert to Protestantism also. This phenomenon led Descartes, as usual, to seek and to enlarge on the *a priori* causes from which it could be deduced; and hence to study astronomy, as well as the particular nature of comets, and also of the light which we receive from all these bodies. These researches were first utilised in his *Monde*, but became known to the world in his *Meteors*, a treatise appended to his *Discourse* as a specimen of the result of his method. So, again, in 1633–34 he made special studies in physiology, and sketched out the treatise *De Homine et de Foetu,* in which he sought to explain *a priori,* as usual, the formation of the human frame. But we must consider his theories on these subjects in their matured form.

HAVING BEEN CHEERED with a visit from his friend Mersenne in 1630, the latter found that his mathematical correspondent Beeckman was claiming to be Descartes' teacher, and even parading the treatise on music which he had received ten years before in manuscript, and had kept and copied, as his own. Whereupon Descartes wrote him a very sharp letter, recalling his treatise, and giving him a very plain opinion on his conduct. This vigorous and outspoken document is still extant.

In the end of the same year the Comte de Marcheville, named French ambassador at the Porte, wished to be accompanied to Constantinople by many learned men, on account of the probable riches in manuscripts and in new observations which might result from a proper investigation of eastern Europe. Gassendi accepted, but was hindered from going at the last moment. Descartes, confessing that a few years earlier nothing would have been more to his mind, declined, as having now finished his travels, and being occupied in more serious work.

In 1632–33, and up to the end of the latter year, he was principally busied with the composition of his *Monde*, a treatise which he had long promised Mersenne, and in which he proposed to explain the *a priori* principles of all physics. This work, which may now be read in the collected editions, is a curious specimen of the over-application of mathematical reasoning to Nature. But such was not Descartes' opinion. Writing to Mersenne to seek from him some description of the known comets, he says:

> Since the last few months I have advanced very far into the heaven, and having satisfied myself concerning its nature, and that of the stars which we see there, and many other things which I dared not even have hoped some years ago, I am now become so bold that I venture to seek the cause of the situation of every fixed star. For though they seem irregularly scattered through the sky, I have yet no doubt that there is among them some natural order which is regular and determinate. The knowledge of this order is the key and foundation of the highest knowledge which men can have concerning material things, because by means of it one may know *a priori* all the various forms and essences of terrestrial bodies; whereas without it we must be content to guess them *a posteriori.*

He adds that it was greatly to be wished that some patient person would write down an exact description of the actual state of the heavens, without hypothesis or conjecture, after the manner of Bacon. This would be the greatest help to the theorist. For all through his life he made his experiments to verify and confirm a hypothesis previously formed. Hence he thinks we should always start from easy and obvious observations, as less likely to deceive us, and only admit the more obscure and delicate when we have a Scientific theory to direct us.

His treatise on the world imagined the Deity beginning with a homogeneous quantity of matter, which was endowed with nothing but its mathematical qualities—extension and mobility—and merely lending His aid to its proper disposition. He proceeds to show how, from such a hypothesis, all the phenomena of the world—the stars, the earth, the animal world, and man—all but his reasonable soul can be explained *a priori.* He dilated specially on the nature of Light, and its propagation through space.

I need not go in greater detail through this treatise, the matter of which was worked up in the *Essays* and *Principles*. There are no doubt happy conjectures in it, and many really large and splendid views. But its principal interest in Descartes' intellectual history is, that it asserted both the movement of the earth and the circulation of the blood. The theory of the circulation of the blood, though it lost Harvey, its discoverer, his practice, did not materially affect Descartes; and he again announced it in his *Discourse on Method.* But the earth's motion was another matter. He had indeed proposed to publish his *Monde* anonymously, in order, he writes to Mersenne, that he might be free to *disclaim it*—a curious confession. But disguises of such a kind were then easily seen through. As he was just about to prepare for the printing of the treatise, and had made inquiries for Galileo's book on the same subject, to compare it with his own, he heard to his astonishment that the doctrine of the earth's motion round a fixed sun had been condemned by the Inquisition, and Galileo sentenced to heavy penalties.

THE CIRCUMSTANCES of Galileo's case were briefly as follows. So far back as 1613, Galileo, having publicly taught (under the protection of the Grand Duke of Tuscany), both in lectures and in conversation, the motion of the earth, published a work concerning *Spots on the Sun* (Rome: 1613—quarto), in which this doctrine was asserted. The cardinals of the Congregation of the Inquisition, having deputed a committee to examine the book, declared it both *absurd and false in philosophy, and erroneous in faith.* But nothing further was then done, owing to the favour in which Galileo stood with Pope Urban VIII. But as Galileo continued to teach his views, it was ordered by the Congregation on the 29th of February 1616 that Cardinal Bellarmine should send for Galileo, and persuade him in a friendly way to renounce his heresy. This was done before witnesses, and, on being admonished, Galileo promised to obey, and was not required to give any other security. Nevertheless, in the following March a formal decree was passed by the Inquisition against this doctrine, which they attributed to Pythagoras; and in it Copernicus, Diego de Zúñiga, and Foscarini, who had supported it, were censured by name, without adding that of Galileo. This decree was received with much ridicule through the foreign learned world, which did not recognise the infallibility of this particular Congregation, and thought them but poor astronomers. It seems that Galileo, when he found this independent attitude assumed throughout educated circles, was ashamed of his own compliance, and sought grounds for withdrawing it. When the Congregation, in 1620, was moved by Scientific opinion to modify its censure of Copernicus, so far as to permit his theory to be argued *as a hypothesis,* Galileo wished to assert the equal freedom of opinion about the theory in Italy. He published, in 1632, his *Dialogues on the System of the World, according to Ptolemy and Copernicus,* and the book was received with public applause, rather as a recantation of the author's submission than a real conforming to the Church. Probably some new and stronger hand presided at the Councils of the Inquisition, for Galileo was forthwith seized, tried, and condemned of heresy, in spite of the permissory decree of 1620. He was told that he "had consequently incurred all the pains and censures of the sacred office, from which, nevertheless he was promised absolution, provided he, with a true heart and unfeigned faith, abjured and renounced before his judges the said errors and heresies" (22nd June 1633). But when he had formally submitted to this decree, and had, with his hand on the Gospels, and on his knees, promised never to say or do anything against this ordinance, he was brought back, with the usual good faith of the Inquisition, to prison; and though released presently on account of his great age and the interest of the Grand Duke, he was condemned to be again shut up in the prison of the Holy Office, and to recite once every week for three years the seven penitential Psalms. He was ordered afterwards to retire to a country house in the territory of Florence, and remain there for the rest of his days. This last decree was dated 23rd August 1634.

News of these proceedings, which excited a great stir through Europe—real concern and fear among orthodox philosophers, ridicule and triumph among Protestants—reached Descartes before the printing of his treatise. He seems to have been quite thunderstruck by the tidings. He had started on his Scientific journeys with the firm determination to enter into no conflict with the Church, and to carry out his system of pure mathematics and physics without ever meddling with matters of faith. He was rudely disillusioned as to the possibility of this severance. He wrote at once (apparently 20th Nov. 1633) to Mersenne to say that he would on no account publish his work—nay, that he had at first resolved *to burn all his papers,* for that he would never prosecute philosophy at the risk of being censured by his Church. He says:

> I could hardly have believed that an Italian, and in favour with the Pope, as I hear, could be considered criminal for nothing else than for seeking to establish the earth's motion, though I know it has formerly been censured by some cardinals. But I thought I had heard that since then it was constantly being taught, even at Rome; and I confess that if the opinion of the earth's movement is false, all the foundations of my philosophy are so also, because it is demonstrated clearly by them. It is so bound up with every part of my treatise, that I could not sever it without making the remainder faulty; and although I consider all my conclusions based on very certain and clear demonstrations, I would not for all the world sustain them against the authority of the Church.

Nor would he even have recourse to the plea that the authority of the Church could only be formally pronounced by a council. But at the end of this long letter creeps in a little sentence which shows the real bent of his mind. After many professions that he would never publish anything, he adds to Mersenne: "Nevertheless, as it would be an act of bad grace if, after many and large promises, I only repaid you in failures, I shall all the same let you see what I have composed as soon as I can; but I still ask you a year's delay to revise and polish it. You have reminded me of Horace's *nonum prematur in annum,* and it is only three since I began this treatise." But when the year had elapsed his fears overcame him, and he wrote to Mersenne saying that he could not induce himself to publish. He had, nevertheless, offered to help a certain ecclesiastic in Paris who was preparing a treatise to maintain the obnoxious doctrine, and only retracted his help from the same feeling of cowardice.

It was not till some time after that he was able to see Galileo's book, which Beeckman lent him from Saturday to Monday, and which he looked through in that time. He found in it so many things identical with his own conclusions, that had it been possible he would have suspected Galileo of having obtained them from him. This pride of originality, and contempt of the books of others, were leading features in Descartes' character. But the difficulty about the earth's motion, which was necessary to his physical system, continued to trouble him. He kept long corresponding with Mersenne on the subject, and even had inquiries made at Rome, apparently through Cardinal Barberini, to know whether this difficulty should prevent the publication of his system.

At last—to conclude our review of this point in his mental history—he discovered a compromise, which, if creditable to his sagacity, is hardly so to his good faith or to his orthodoxy. In his *Principles*, published ten years later, he formally denies that the earth moves, but holds that it is carried along, together with its surrounding water and air, in one of those larger motions of the celestial ether, which produce daily and yearly revolutions of the solar system. The earth indeed did not move, but it was like a passenger in a vessel, who, though he were stationary, and properly said to be at rest, is nevertheless carried along in the motion of the larger system which surrounds him. He makes the most of this distinction, as opposing him to Galileo and to Copernicus, and thinks that he may rather be called a disciple of Tycho Brahe. "If we are not to follow either of these systems, we must return to that of Ptolemy, which I cannot believe the Church will ever force us to do, seeing that it is manifestly opposed to experience." Nothing pleased him more than that many Protestant divines in Holland attacked the new doctrine. He thought it might dispose the Catholic Church to accept it.

In 1635 we find him making observations on snow-crystals, which formed the sixth section in his treatise on *Meteors*, and with which he was peculiarly well satisfied. In 1636 he wrote for M. de Zuytlichem, father of the well-known Huyghens, a short treatise on *Mechanics*, and made speculations on the coronae which sometimes appear to us round the flames of candles, and in which the colours do not agree with those of luminous circles round the sun or moon. Meanwhile Renéri had been appointed Professor of Philosophy at the rising University of Utrecht, and thus Descartes found an opportunity of disseminating his views through an official teacher from an important chair. The teaching of Renéri at Utrecht had indeed momentous consequences for his peace of mind in after-years. But for the present the prudence and caution of the new professor avoided all conflict with his colleagues, and with the still established peripatetic philosophy.

5

THE ESSAYS—CORRESPONDENCE AND CONTROVERSIES

1637-41

We are not informed of any particular reasons which made Descartes, some time in 1636, change his determination of printing nothing, and announce to Mersenne that he was about to become an author. Probably this had been his settled intention for many years, and it was only delayed by his scare about Galileo, from which a few years' reflection began to rescue him. The Elzevirs of Amsterdam had made him offers of publishing for him, and so had many Parisian publishers in former years; but when at Amsterdam, and with his book nearly ready, he seems to have taken some offence at the want of anxiety about it shown (no doubt for commercial reasons) by the Elzevirs. Hence he sent a specimen manuscript copy to Mersenne, and consulted him about the convenience of publishing in Paris, as well as of obtaining a copyright privilege from the King of France. The arrangement he proposed was that he should have the choice of paper, form, etc., and should receive 200 copies as his share from the publisher, who was protected by the Privilege from piracy. I cannot find that any money transaction ever took place between Descartes and his publishers. He distributed a large number of copies, but received no other profit. They, on the other hand, annoyed him with complaints that his books had no sales, and that they had lost by the venture. These complaints, of which we hear especially some ten years later, vexed him exceedingly, but chiefly on account of the evidence that the public did not read his books.

The excellent but somewhat fussy Father Mersenne was so delighted with his friend's new commission, that he quite overdid his duty. He immediately showed the precious and long-expected work to many friends, thus disclosing the author, who had meant to remain secret; and moreover, he obtained from the king an extensive and pompous privilege, naming Descartes, and permitting him to print not only these treatises, but any he should wish in future, both within and without the kingdom.

THE TITLE of the manuscript sent to Mersenne was as follows: *The project of a universal Science which can raise our nature to its highest degree of perfection. Then the Dioptric, Meteors, and Geometry, in which the most curious matters which the author could select to show evidence for his proposed universal Science are explained in such manner that even those who have not studied can understand them.* Descartes was annoyed at Mersenne's officiousness; and finally perhaps on account of the difficulty of correcting his proofs, and still more his plates, at a distance came to Leyden, and put his manuscript into the hands of John Maire (or le Maire), who proceeded to print it. The Privilege from the King of France delayed the last sheet in waiting for its insertion; and when it came, Descartes thought it so pretentious, especially in promising other books, that he only printed an extract, omitting mention of his own name. The book was announced as in the press early in 1636, and its delay caused much impatience at Paris, the *Dioptric* being nearly a year in printing; but I suspect Descartes was chiefly to blame. A set of sheets as they were struck off were procured from the publisher by M. Beaugrand, who thus was informed of the progress of the book.

So it appeared, a moderate quarto, not very well printed, and on poor paper, the 8th June 1637. Its publication marks an epoch not only in the philosopher's life, but in the history of human thought. It bore the more modest title: *Discourse on the Method of properly guiding the Reason in the research of Truth in the Sciences. Also the Dioptric, the Meteors, and the Geometry, which are Essays in this Method.* We know from various letters that notwithstanding the long delay in publication, and the maturity of Descartes' views, certain portions of the book were afterthoughts to its general plan, and were hurriedly written even while it was going through the press. This is the case (1) with the article on the proof of the existence of the Deity, the obscurity of which he excuses partly from its being an afterthought and composed in haste, partly from the curious reason that to explain properly the certainty of his demonstration, he should have dilated more fully on the uncertainty of our knowledge of material things, and thus have set forth sceptical arguments which he thought inexpedient for the public whom he desired to instruct. (2) His Geometry was also an afterthought, written, and even partly invented, while the *Meteors* were in the press, and was even designedly so composed as to be very difficult, and only understood by mathematicians of a high order. He omitted what was obvious, and in the solving of problems only gave the means of solution, and not each step in the demonstrations. He even chuckles in his letters at the number of professed mathematicians who were unable to follow his arguments, and tells us that not a single professor in the new universities of Holland was able to open his mouth upon the subject. As soon as he had determined to print this tract, he had twelve copies struck off on special paper and handsomely bound for such special scholars as Mersenne should choose to read it in France. The omission by some accident of M. de Roberval, a leading mathematician in Paris, in the distribution of this tract and of the *Essays*, is said to have earned Descartes his lifelong enmity and much bitter controversy.

It should be added that though he gave in the title and divisions of his work three applications of the *Method*—one in mathematics, which he considered the most important, and as demonstrating what the other applications only illustrated; a second *(Dioptric)* in mathematical physics; a third *(Meteors)* in physics proper—there was a fourth and a most important illustration in physiology embodied in the opening discourse. It was the theory of the circulation of the blood. Although he could not refrain from claiming originality here also, and denying that anyone else had explained the motion of the heart in the way he now proposed, yet he here and here only, perhaps, in all his works concedes to another the honour of a great discovery, and tells us that Harvey deserved the credit of being his pioneer.

He also gives a pretty full sketch of his general physics, as set forth in the treatise he had suppressed; and regarding this, as he himself tells us in a letter, "I should not have been so imprudent as to speak of it in the way I did, if I did not desire to let it see the light, should the world wish it, and thus make it worth my while, and safe. But I may tell you that all I have printed is merely to prepare the way for it and sound the ford."

HAVING NOW GIVEN ALL the external circumstances which are of interest in the publication of this famous volume, we will say a word on the general value and import of its contents. We may dispose of the latter *Essays* very briefly. Though the work was written in French, and thus appealed to the ordinary public—a bold innovation in itself. These Scientific inquiries were dry and difficult, nor did any but a very select public ever care to study them. And now they have been long superseded and surpassed by the advance of Science, so that they are even omitted in most modern editions of the *Discourse on Method.* The *Geometry,* though acknowledged as an epoch in pure Science and still reverenced by all the mathematicians who never have seen it, has been absorbed in later works. The *Dioptric* has been forgotten in the light of Newton's optics, and the *Meteors* have been laid aside in favour of the purely experimental method of later and less celebrated investigators.

But it must never be forgotten that Descartes' treatises, though full of assumptions and defaced with curious *a priori* absurdities, nevertheless insisted upon the only true and solid path ever followed by physical Science—its reduction to the mathematical laws of figure and of motion. Having first shown, by the earliest of all his discoveries, that all problems in figure could be reduced to arithmetical formulas, and that these could be generalised by the use of algebraic symbols, he insisted that nothing should be assumed in explaining the laws of nature but the laws of figure and motion. He cast to the winds the whole apparatus of occult qualities, intentional species, and other assumed secrets by which the Scholastic Peripatetics endeavoured to explain—and by which they succeeded in obscuring and confusing—nature. We, to whom these Scholastic theories are things long past, cannot now feel the novelty and the boldness of Descartes' conception, that all nature can be represented in algebraic formulae, and its laws expressed in definite equations. Nor was he bound to show more against the schools than this, that his method was simpler and more satisfactory than theirs in explaining nature. For as theirs was purely hypothetical, his need only be a simpler hypothesis. He banished all hidden mysteries, all occult forces from the world. The only difficulties to be encountered were the difficulty of choosing between various *a priori* explanations, which were equally satisfactory, and that of the complication of the various forces or elements which entered into the solution of the problem. The former was met by the making of experiments which must be so constructed as to exclude the plurality of causes, and verify the true *a priori* deduction. Thus Descartes recognises both the method of agreement and the method of difference in our modern inductive logic. The complication might often be such that any solution became impossible, and in his mathematics he particularly prided himself on discovering the conditions of problems incapable of solution. He felt, like all the 16th and 17th century reformers, waste of time on sterile speculations to be one of the greatest of human ills, and desired above all things to render Science subservient to the arts and to the comfort of mankind.

BUT IF THE advance of human knowledge has consigned to oblivion these splendid essays in mathematics and in physics, the *Discourse on Method* was too remarkable a manifesto—too clearly the trumpet-note for the resurrection of the human mind from the death of formalism—ever to be forgotten or to grow old. Written in the popular tongue of Europe, and with a clearness and simplicity rarely equalled even in French prose, the best prose in modern Europe, it produced an electric shock throughout the learned world, which no other work of the kind ever did in the history of philosophy. The other forerunners and anticipators of modern thought are mentioned in histories, but who ever reads Bruno, or Campanella, or Vanini? Bacon indeed alone still survives, but is read only for his style, not as a real constructor of a new system of knowledge. But even impostors like Chandoux traded on the recognised necessity of abolishing the Aristotle of the schools; the real problem was to substitute the something new for the old which was laid aside. Now in all his positive attempts Bacon had signally failed. He had so little instinct for discoveries that he could not recognise them even when they were made, and he ridiculed as inconceivable the theory of the earth's motion. Still worse, he wanted the training which is absolutely necessary to the discoverer in physical Science; he was ignorant of mathematics, both pure and applied. He was not regarded by his immediate successors as of any real importance. Descartes, Hobbes, Newton, Locke hardly mention him. It was probably through Gassendi's influence that he afterwards became fashionable. But Descartes did not publish advice on how to make discoveries till he could show the actual discoveries he had made, and this again distinguishes his *Discourse on Method* from the *Instauratio Magna* of his eminent contemporary. The closer analysis of the *Method* will occupy us when we proceed to give a summary of his whole philosophic system.

The book being published, its author began to await with curiosity the criticisms of the learned, which he expected in manuscript, as the habit of reviewing by means of a periodical press had not yet arisen. He did not feel the smallest fear that he might be refuted; for though he writes to his old Jesuit tutors with affected humility, "If you will take the pains of reading the book, or making those of the company who have leisure to read it—and if after having noted the faults, which will doubtless be found in it in great numbers, you will do me the favour of informing me of them, and so continue to teach me—I will owe you a great obligation, and do my best to correct them according to your good instruction." He says to more intimate friends that he "does not believe there are three lines in the book which can be rejected or changed; and that if there be the least falsehood in any the least part of what he had published, his whole philosophy was not worth a straw."

THE THREE YEARS following the publication of the Essays were occupied in controversies almost completely mathematical concerning the principles of the *Geometry* and *Dioptric*; nor do we hear much of attacks on the *Discourse* or on the *Meteors*. A few obscure persons said something about the demonstration of the existence of the Deity; but Descartes held them in too much contempt to reply to them. "I may tell you that I should not be less ashamed to write against a man of that sort (M. le Sieur N., backed by the Capuchins), than to stop and pursue any little cur that comes barking after me in the street." Also on Petit, he says: "I think no more of him than I do of the abuse given me by a parrot hanging in a window as I pass in the street."

He solicited everywhere objections, provided always the objectors consented to have their letters printed together with his replies, intending to make a volume of these letters. This intention he never completed; but in his collected letters we find all the materials preserved.

His first objectors, or critics, were Dutch and Belgian savants; and with some of these he was even offended that they ventured to offer an opinion so hastily on a work which he considered should require long and careful study to appreciate it. Ciermans, however, a Jesuit of Louvain, made acute objections to the theory of light contained in both *Optics* and *Meteors*; and Morin of Paris followed up the question with an elaborate discussion on the theory of light.

The most serious criticism of his *Dioptric* and *Geometry* came from Fermat, who not only wrote objections to the former, but sent Descartes his own treatise, *De Maximis et Minimis*—to show that there were important omissions in the *Geometry*. Descartes defended himself, and even attacked Fermat's treatise, seeking to prove that his demonstration of the finding the maximum tangent to a parabola was only an accidental proof, and that Descartes himself had stated the true special conditions of the problem. This controversy, in which Roberval of Paris supported Fermat, and others stood by Descartes, lasted a long time, and was only stopped by the fatigue of both principals. It ended, however, with letters of mutual respect and friendship, of which Descartes to Fermat are very graceful. "I had no less pleasure," he says, "in receiving the letter which does me the favour of promising your friendship, than if it came from a mistress whose good graces I had passionately desired; and your other writings preceding it remind me of the Bradamante of our poets, who would receive none as her servitor who had not first proved himself against her in combat. Not that I pretend to compare myself to that Roger who alone in the world was able to resist her; but such as I am, I assure you I highly esteem your merits." The historians of mathematics are, however, of the opinion that Descartes was wrong; and if so, he showed very censurable obstinacy, and even unfairness, in his conduct through the controversy.

He also busied himself for a time on the problem of the Cycloid, which Roberval had proposed, and on which Pascal composed a famous tract under the title of the *Sieur d Ettonville,* giving a history of the discussions upon it. Very honourable mention is made here of

Christopher Wren, then a student at Wadham College, Oxford. But Descartes was getting tired of mathematical problems, and after contributing to the question turned from it.

We hear of no writings (beyond letters) in this period, except the tract on *Mechanics*, which he only sent to Zuytlichem in 1638, but had composed earlier; and a tract on the Geostatic Question, in criticism of a worthless book of M. Beaugrand, which his friends entreated him to print; but he refused. He had, in fact, written it so hurriedly, that he had fallen asleep as he reached the close, and had it copied and sent off to Paris the next day without reading it over—a curious insight into his habits in writing Scientific letters. It is believed that this tract survives in a long letter on the question of whether a body is heavier near or far from the earth's centre.

HIS CORRESPONDENTS in Paris were much concerned when they heard that he was about to abandon the study of mathematics; but his reply to the protest of Des Argues is important in understanding his philosophy. "I have only resolved," he says, "(now that my hair is turning grey) to give up abstract geometry—viz., questions which test mere ingenuity—and will confine myself to the pursuit of another geometry, which proposes to explain the phenomena of Nature. In fact, Des Argues would find that *all my physics were nothing but geometry* if he took pains to read the *Meteors.*" Among the various commentaries offered by friends to facilitate the *Geometry*, that of M. de Beaime appeared to Descartes the most thoroughly satisfactory.

He had, moreover, trained a French Huguenot valet named Gillot in his analytic geometry, so as to be better able to explain it than anyone else. This valet, who must have been a clever youth, was sent abroad with Descartes by his parents to avoid Roman Catholic pressure, had learned Flemish and some Latin—nay, even (by residence in England) English, which Descartes never knew—and was on such terms with his master that Descartes tells us he treated him *en camarade*; and when mathematical problems of lesser import were sent him, he used to say, *give that to Gillot.* He took pains to have this young man, who turned private teacher of mathematics, established as secretary to Sainte Croix, reminding him that he had been treated with such confidence and familiarity, and had associated so much of late years with people above him in quality, that he could not be expected to perform the menial services of a Parisian valet. He praises his fidelity, and says his parents had brought him back from England, fearing for his moral principles when out of control. He was able to earn four or five hundred *écus* per annum by teaching mathematics at La Haye or Leyden.

THE MAIN EVENTS in the outward life of our philosopher during the years 1638–41, in the latter of which his second great work, the *Meditations*, appeared, are to be found in connection with the University of Utrecht, where his disciple Renéri had been preaching his doctrines. But Renéri was in bad health, and he died tragically enough at the age of 45, on the very day of his wedding. His friends had persuaded him to marry, in the hope that better care of his home and comforts might improve a constitution worn out with hard work and much meditation. But he was taken ill at the wedding-feast and died in a few hours.

This man was the specimen of the enthusiastic Cartesian pupils who presently thronged the universities of Holland. He writes to Mersenne, *"Is est mea lux, meus sol; ille mihi semper Deus";* and when he died, and another professor, Æmilius, was ordered by the chief magistrate of Utrecht to pronounce an *éloge* on the deceased and on the new philosophy. We hear of similar extravagances, also expressed in Latin: Descartes was called "the only Archimedes of the age, the true Atlas of the universe, the Hercules, the Ulysses, the Daedalus, etc., etc., of Science." It is true that Latin always admitted these follies, and that many men to the present day would write compliments in Latin of which they would be ashamed when expressed in their own tongue. But these phrases are not in any degree stronger than the English of William Molyneux in the preface to the *Meditations*, which he translated in Dublin and published in 1680.

> In fine, such is the Excellence of these six Meditations, that I cannot resemble his Performance herein better than to the Six Days Work of the Supream Architect; and certainly next to the Creation of All things out of Nothing, the Restoration of Truth out of Errors is the most Divine Work; so that (with Reverence be it spoken) the Incomparable Descartes does hereby deserve as it were the name of a Creator. In the first Meditation we are presented with a Rude and Indigested Chaos of Errors and Doubts, till the Divine spirit of the Noble Descartes (pardon the Boldness of the Expression) moves upon the confused face of these Waters, and thereout produces some clear and distinct Light; by which sun-shine he proceeds to bring forth and cherish other Branches of Truth; Till at last by a six Days' Labour he Establishes the Fair Fabrick (as I may call it) of the Intellectual World on foundations that shall never be shaken. Then sitting down with rest and satisfaction he looks upon this his Off-spring, and Pronounces it Good.

But though Renéri died so inopportunely, he had sown dragon's teeth in converting a young physician named Regius (Le Roy), who took up the tenets of Descartes with ardour, and was appointed assistant professor of botany and medicine shortly after Renéri's death.

He had sought anxiously for an introduction from Renéri to the great philosopher, and addressed him in terms of flattery as the true cause of his promotion to the chair. Descartes replied with courtesy, and with apparent modesty, to this and the other demonstrations of enthusiasm on the part of his Utrecht adorers, and affected to think little of the important consequences of his teaching. But what proves to the more careful inquirer that he was really overcome by their attentions, and even to some extent puffed up by them to undue vanity, is that he consented to abandon his retreat at Egmond, and come to live near Utrecht, helping them with his advice, and even correcting the form of the theses with which they attacked the old philosophy. Still stronger evidence is afforded by the fact that at the moment when Regius began to natter him about the physiological side of his physics, and its importance to medicine, he begins his only foolish and un-Scientific project of indefinitely increasing by his philosophy the length of human life.

> I have never been so careful to preserve myself as now; and whereas I formerly thought that death could not deprive me of more than thirty or forty years, it cannot now surprise me without robbing me of the hope of at least a century. For I think I can see very evidently, that if we merely avoid certain faults which we ordinarily commit in the regimen of our life, we might without any other invention arrive at a far longer and happier old age than we do at present. But as I require much time and many experiences to examine all that belongs to the subject, I am now working at an Abridgement of Medicine, which I take partly from books, partly from my reasoning. I hope by this means to obtain some delay from Nature, and so pursue my designs better in the course of years.

So, also, we are told that Sir Kenelm Digby (whom Baillet calls Lord Kenelm, Count d'Igby, Knight of the Garter, etc.) went to see Descartes in Holland, and besought him to use his great knowledge of the human body above all things to prolong human life. Descartes told him he had already thought much about it, and dared not promise to make man immortal, but that he felt sure he could prolong his life beyond that of the patriarchs. This was communicated by St. Evremont to his biographer Desmaiseaux, with the remark that the matter was well known in Holland. These notions so impressed his followers that some of them thought him almost immortal; and it is said that his friend the Abbé Picot for a long time refused to believe the news of his death. He himself soon recovered from these vagaries, and told his friends that if he had found no means of prolonging life, he had at least succeeded in overcoming all fear of death.

IT IS to be noticed that even among the Protestant divines there were some who affected the new philosophy, in particular Heide (Heidanus), an eminent orator, who preached *à la cartesienne* and attracted great attention. But of all the rest, though many pretended him friendship, Descartes writes that not one of them was his friend. In fact, the spirit of free inquiry, and the rejection of all received dogmas, though somewhat more consistent with the theory of Protestantism than with that of the Church of Rome, was not in the least more pleasing to Evangelical divines. For to these persons the authority of their synods or their Bible is just as absolute as that of the Church to the Romanist, nor are there any men more intolerant of opposition. Thus a party of theological opponents to Descartes' philosophy soon was formed, and the man who came to lead it was a celebrated controversialist and preacher, Gisbert Voët. We shall take up this controversy later, when it became clearly pronounced, and occupied the philosopher's full attention. At first the dispute was confined to the angry discussion of theses in physiology between Aristotelians and Cartesians, in which the circulation of the blood formed the principal ground of contention.

Descartes was rather pleased than otherwise to find opposition from Protestant theologians, for he hoped it would reconcile the opposite party to his views. Indeed he did not hold even Roman Catholic priests in good esteem; for in writing to recommend two of them, who lived near him at Alkmaar, and who delighted him with their music, to the attention of Zuytlichem, and to his influence with the Prince of Orange, he apologises to his friend for knowing them, and says that he keeps clear of the frock wherever he goes. This anti-theological bias is carefully to be noted in estimating the character of a man who was so profuse in his declaration of adhesion to the Church.

BUT HOWEVER HE may have despised the regular clerk of all denominations, it was a different thing with the learned orders, which contained most of his friends, and especially the powerful Order of the Jesuits. He had sent them his book, praying their favour and support, and he was, above all, anxious that they should not declare themselves against him. Hence his courteous and deferential reply to Ciermans of Louvain, as he hoped that the adhesion of one would imply that of the rest, so great was the inter-connection and union among all the members of the Order.

Accordingly he was much surprised to learn that Bourdin, who was Professor of Mathematics at the College of Clermont, had set theses to be discussed in which the preliminary statement by the Professor was an attack on his philosophy. He wrote at once to the Rector, complaining that Bourdin's objections had not been sent him privately for the purpose of a reply, and besought that he would do him justice. The Rector replied by directing Bourdin to explain the matter to Descartes, and declared it merely a private dispute with a single member of the Order. But Descartes, before receiving this answer, was much excited, and told his friends he was going to battle against the whole Order, for he could not believe that they would tolerate a private dispute. Nor did the Rector's answer undeceive him. He had evidently received a strong impression of the unity and deliberation of all Jesuit action at La Flèche. He therefore began to prepare for the struggle; and as Bourdin argued on Scholastic grounds, he designed a formal refutation of the Scholastic philosophy, following section by section the course of one of their best handbooks. He accordingly sent for books written by the Order or in use among them, and turned to the school logic which he had abandoned more than twenty years earlier. Neither he nor Mersenne knew even the names of these books, beyond that of the *Conimbricenses*, or handbook written at Coimbra in Portugal; but he picked up at Leyden the handbook of philosophy by the Cistercian Eustache, called Le Feuillant, from the convent at Languedoc, which he thought most suitable to print with a running refutation at the end of each section. But he found that brother Eustache was still alive, which caused him difficulty, and that the *Conimbricenses* were too long; so he postponed the work, though full of it, and eager to finish it, till after the appearance of his forthcoming *Meditations*. These events took place in the latter half of 1640.

~~JESUS AND THE WOMAN TAKEN IN ADULTERY~~

DESCARTES AND THE WOMAN CAUGHT STUDYING PHILOSOPHY

Descartes stooped down, and with his finger wrote on the ground ...
and said unto them, "He among you who can speak 14 languages,
let him first cast aspersions on her."

We hear of the friendship of Descartes at this time with many notabilities of Holland, and among others, with De Saumaise (Salmasius), who was, like him, a French *gentilhomme* settled abroad, but a Protestant. This Salmasius, though not a professed philosopher, proclaimed himself a Cartesian, and was as friendly to Descartes as the latter—being afraid of his exceedingly quarrelsome and jealous nature—would permit. In fact, the very suspicion of being a friend of his rival Heinsius was enough to excite Salmasius' hostility. At the Hague, Descartes was not only intimate with Zuytlichem and other gentlemen about the Court of the Prince of Orange, but with the Court of the ex-King of Poland.

At this time his most distinguished lady-friend, the Princess Elisabeth, was still very young, and the great female *savant* of Holland was Anna Maria Schurman of Utrecht, of whom we hear the most wonderful accounts. She was now (1640) twenty-eight years old, and was said to be perfectly skilled in all the tongues of Europe, not excepting Turkish, as well as Greek and Latin, Hebrew, Syriac, and Arabic. Her French style was highly admired by Salmasius. She was, moreover, devoted to the fine arts, and had produced the most exquisite workmanship in painting, illuminating, sculpture, and engraving on wood, marble, and precious stones. She understood mathematics, dialectics, the Scholastic philosophy, and, above all, theology, being versed in St. Thomas and the Holy Scriptures. With all these accomplishments, she was of a deeply pious and modest mind, given to prayer and study, avoiding all worldly delights, adopting a vow of virginity with the touching motto of St. Ignatius, *amor meus crucifixus est.* This is admitted even by the Catholic Baillet. But her enthusiasm led her into the extremes of Protestant sectarianism; she adopted the newest and extremest Reformations, and, as Descartes tells us, having come under the influence of Voët, turned to controversial theology.

Descartes knew this lady, for he proposed to attend one of the disputations on his philosophy held in Utrecht, if he were concealed in the *écoute* or curtained seat of Mdlle. Schurman. But he evidently thought nothing of her deeper Science—which has no doubt been as much exaggerated as the philosophy of the Princess Elisabeth—for he says that she spoiled a remarkable talent for the fine arts by learning Calvinistic controversy from Voët. In M. Foucher de Careil's recent book on Descartes and the princess, we have this interesting fact, that Mdlle. Schurman came to have a horror of Descartes, which we should think natural enough in a pious pupil of Voët; but there is added the anecdote, that he once, on visiting her, found her occupied with the study of the Hebrew Scriptures, and expressed his wonder that so clever a person should be occupied with such trivialities. Upon her protesting that no study could be more profitable, he told her he too had once been very curious to know exactly what Moses had said on the subject of the creation, and had even learned Hebrew to judge for himself in the original, but finding that Moses had said nothing *clare et distincte,* he had laid him aside as affording no light in philosophy. She afterwards used to thank God she had escaped from the influence of so profane and impious a person. It was no doubt this spirit which alarmed orthodox Protestants like Voët, and caused them to see in Descartes not only a disguised Jesuit but an atheist.

Another important friend at this period was Lord Charles Cavendish, brother of the Marquis of Newcastle, who was exceedingly struck with the new philosophy, and coming to know Mersenne, begged him to send anything Descartes should write, which the good Father did, pretending to do it without the philosopher's knowledge or consent, in order to enhance the compliment. This Lord Charles Cavendish sought to influence King Charles I to send for Descartes and for Mydorge, as a practical optician, and give them both comfortable posts in England, in order to advance Science there. They even offered to provide for Mydorge's family. Descartes was not unwilling to go, as the Utrecht troubles were increasing; but the outbreak of the civil war in England stopped the negotiations. There is no evidence of the assertion frequently made that the King of France made a similar offer.

6

THE MEDITATIONS, AND THE OBJECTIONS OF THE LEARNED

1641

We now come to the publication of the second of Descartes' works, and that for which he himself in intimate circles often expressed his own preference, as by far the most original and profound. It was composed in Latin, and printed at Paris in octavo, by Michel Soly, where it appeared in August 1641, with the king's privilege, and the approbation of the doctors of the Sorbonne, under the title *Meditations concerning the First Philosophy, in which are demonstrated the Existence of God, and the Immortality of the Soul.* Descartes wrote to Mersenne:

I have no intention of ever printing my *Principles* or the rest of my *Physics*, or anything else except three or four sheets concerning the existence of God, to which I feel bound in conscience. As regards the rest, I know no law obliging me to give to the world things which it shows are not desired. For every twenty approvers who would do me any good, there would be thousands of ill-disposed people who would injure me whenever

> they could. This is what experience has taught me during the last three years; and though I do not repent me of what I have printed, I have so little fancy to repeat it, that I will not even have it printed in Latin, so far as I can prevent it.

In fact, we find the philosopher in very bad humour with the public, and very angry at all their objections, though he had himself challenged them. What seems to have most hurt his vanity, which was excessive, was that though he had sketched out in the *Discourse on Method* the main arguments of his *World*, the suppressed treatise, and had spoken vaguely, and refused to explain its details, in the hope that the public would cry out for it, there had been no such demand. People had found ample materials for comment and criticism in the published work; the Utrecht theologians had even found grave hints of scepticism, or atheism, as they called it, in various parts of his argument, and rather deprecated further works of the same kind. Hence he was determined to give to the public first of all the metaphysical and religious side of his system, before formally attacking the Scholastic philosophy.

DESCARTES MOTIVATES HIS DISCIPLES

Before attempting publication, Descartes spoke to his protégés:

"Do not let your minds be disturbed. You believe in Science; believe also in me. The House of Knowledge has an insecure foundation; if that were not so, would I have told you that I am tearing it down to prepare a foundation for real Knowledge? And if I go and prepare a foundation for Knowledge, I will come back and take you to study with me, that you also may know what I know. You know the way to the house I am building."

DESCARTES AND THE WAY TO CERTAINTY

Blaise Pascal said to him, "Magister, we don't know if we exist, so how can we know anything for certain?"

Descartes answered, "*Cogito ergo sum.* No one comes into existence except through the mind. If you really know your own mind, you will know of your existence as well. From now on, you do know you exist and have found true certainty."

Henricus Regius said, "Magister, show us Science, and that will be enough for us."

Descartes answered: "Don't you know me, Regius, even after I have been with you such a long time? Anyone who has heard me has found Science. How can you say, 'Show us Science'? Don't you believe that I am in Science, and that Science is in me? The words I say to you I do not speak on my own authority. Rather, it is the Science, discovered by me, that is doing this work. Believe me when I say that I am in Science and Science is in me; or at least believe on the evidence of the works themselves. Very truly I tell you, whoever follows my method will do the works I have been doing, and they will do even greater things than these, because I have shown the way to Science. And you will know whatever you question using my method, so that certainty may be established through me. You may question anything with my method, and you will know the answer.

DESCARTES PROMISES THE SPIRIT OF PROGRESS

"If you respect me, read my books. And I will pursue Science, and Science will give you the *primum movens* to help you and be with you forever—the Spirit of Progress. The world cannot accept it, because the world neither sees Progress nor knows Progress. But you know Progress, for it accompanies you and inspires you. I will not leave you as independent scholars; I will mentor you. Before long, the world will not hear me anymore, but you will hear me. Because I know, you also will know. On that day you will realise that I have found Science, and you have found me, and I have found you. Whoever has my books and reads them is the one who respects me. The one who respects me will be edified by Science, and I too will edify them and reveal my philosophy to them."

Then Marin Mersenne said, "But, Magister, why do you intend to reveal your philosophy to us and not to the world?"

Descartes replied, "Anyone who respects me will follow my teaching. Science will edify them, and it will come to them and make its home with them. Anyone who does not respect me will not follow my teaching. These words you hear are not my own; they belong to the Science I have discovered.

"All this I have spoken while still with you. But the *primum movens,* the Spirit of Progress, which Science will send in my name, will enable all things and will remind you of everything I have said to you. Rationalism I leave with you; my skepticism I give you. I do not deceive you as the world deceives. Do not let your minds be disturbed and do not be misled.

"You heard me say, 'I am going away and I am coming back to you.' If you respected me, you would be glad that I have discovered Science, for Science is greater than I. I have told you now before it happens, so that when it does happen you will believe. I will not say much more to you, for the Evil Demon is coming. He has no hold over me, but he comes so that the world may not discover Science and learn exactly what Science has revealed to me.

"Come now; let us doubt."

As he considered the subject so full of difficulty, and liable to so much misprision, Descartes determined to send the manuscript to Mersenne with a letter to the doctors of the Sorbonne, or Theological Faculty of Paris, praying their approval and sanction. He also requested him either to have a dozen copies printed, or to have manuscript copies sent to the ablest theologians and philosophers he could find, in order that they might send him their criticisms, and that he might print them together with his replies at the end of his treatise. He made the same request of his Dutch friend Zuytlichem. But Mersenne pointed out to him the risk of printing any copies privately, especially in Holland, as the printer would be sure to let them out, and the Protestant theologians would attack them before the book, with its Catholic support, could be published. Hence he had manuscript copies distributed by Mersenne; and though he was offended on the one hand with the promptness of Caterus, a Louvain theologian, in framing his objections to so profound a work, yet being impatient on the other that Mersenne should collect as many as would fill up a good volume, he soon succeeded in collecting and reducing to order a considerable body of criticism.

The objections were divided into six heads: the first by Caterus; the second a collection from various hands by Mersenne; the third by Hobbes; the fourth by Arnauld; the fifth by Gassendi; and the sixth (and last in this edition) a collection similar to the second. The seventh objections, by the Jesuit Bourdin, which were not written in time for the volume, appeared in the second (Latin) edition of 1642, which Descartes brought out when he heard that in any case the Dutch publishers would pirate it. He substituted in the title *The Distinction of Soul and Body* for the *Immortality of the Soul,* seeing that this had been pointed out to him to be a misnomer, as he only discussed the immateriality but not the immortality of the soul throughout the *Meditations*. He also added a letter to Father Dinet, Provincial of the Jesuits, at the close of the objections. This edition is, accordingly, much more complete and valuable than the *princeps* of Michel Soly.

The work itself is merely an expansion of the metaphysical side of the *Method*, omitting all Scientific applications, and therefore occupies similar but far narrower ground. We will therefore give the slightest sketch of it here, as we must return to it in considering Descartes' philosophy. But we may dispose of the objections, as most of them have merely a historic interest, and only served to elicit from him more explicit statements of his views.

THE WHOLE OBJECT of the six Meditations is to prove that the knowledge of God and of the mind (or soul) is the most certain of all knowledge; and this is done in two ways: first, by showing the uncertainty of all our knowledge of bodies; and second, by offering demonstrations for the existence of ourselves (as thinking beings) and of the Deity. The first Meditation establishes the general doubt as to all our previous knowledge, and especially that of the senses, which we already know from the *Discourse on Method.* The second repeats the argument establishing, by the very fact of doubt, our own existence as beings who doubt. It proceeds to show that even external objects cannot be known with any certainty until they become likewise objects of our thought. The third gives us the demonstration of the existence of God, based on the distinction of the natural light, which reveals to us innate ideas, from the blind impulse which leads us to hold the false for the true. Descartes insists that the omnipotent Deity is required, not merely for the creation, but for the conservation, at every moment, of the universe. The fourth, starting from the principle that *whatever we know clearly and distinctly must be true,* proceeds to explain the existence and nature of intellectual (not moral) error from the disproportion between our unlimited will and our limited understanding, the will being a necessary element in belief, and thus urging us to accept as true what our understanding has not grasped. The fifth repeats the demonstration of the existence of the Deity under a new form, and proceeds to show that the truth and reality of the material world are derived from our knowledge of God's existence and His perfection, for not even the strictest demonstration should convince us until we know that we are not the playthings of a superior Being who chose to deceive us. In the sixth is set forth the difference between the action of the understanding and of the imagination; the fundamental difference and yet close union of the mind and body are described, and the errors of the senses exposed. Then follow the evidences of the existence of a material world, including our own bodies, with the object of showing that those things which nobody doubts are not known on evidence so certain and immediate as that for the soul and the Deity.

Descartes himself added an abridgement of these *Meditations* by way of preface, which is not very satisfactory, and omits the mention of many subjects of interest in the fuller discussion. But feeling that he had only shown the immateriality and not the immortality of the soul in his text, he appends something additional to the sketch of the second Meditation. He says that the strict or geometrical proof of this immortality requires an explanation of his whole physical system, but the heads of the proof are found in the Meditations—viz. (1) The first condition of such a proof is to form a perfectly clear notion of the mind as distinct from the body (6th Med.); (2) That what we thus clearly conceive must be true (4th Med.); (3) A clear conception of our corporeal nature (2nd, 5th, and 6th Med.); (4) That substances thus clearly conceived as distinct must be really so (6th Med.), and that the body being thought as divisible, the mind can only be conceived as indivisible (6th Med.) His physics would show that any substance once created was imperishable, except God chose to annihilate it, and that this was the case even with the elements of the human body, which were "discerpted" by death, and the body, as a composite only, destroyed. But the mind being not so composed, would continue to exist according to the natural and universal law of the indestructibility of substance.

By far the best compendium of the work is the fourth section of the *Discourse on Method,* which clearly shows that the whole argument, and the very title *Meditations*, were already then in his mind. He does not appear to have improved his statement in the fuller expansion. As he was growing older, and more worshipped by his French and Dutch admirers, he became more dogmatic; and having once established that whatever he conceived clearly and distinctly was true, and that this clearness often came only after long and close reflection, he began to persuade himself that all manner of metaphysical principles were true because he was persuaded of them. In fact, he failed to notice that the *impetus spontaneus,* which he evidently distinguished from *lumen naturale* —as being unreflective and, as it were, instinctive—might also arise from over-subtlety, and be (as Bacon has it) a *phantom of the cave* as well as a *phantom of the marketplace.*

THE FIRST OBJECTIONS OF CATERUS, made in the friendliest spirit, touch all the metaphysical weaknesses of the argument, especially the relation of the idea of God in our minds to its cause. Descartes had argued that the perfections existing objectively (as he called it) in the idea, must come from a cause in which they existed formally. He also argued that whatever imperfect or false notions we had were derived from non-being, even though they had an "objective" existence in our minds. In all these points Caterus shows the defects and insufficiencies of his argument, and the reply of Descartes merely proves that he had made up his mind to stick by his position, and defend it at all hazards. He shows, however, that he had avoided the argument which runs back from effect to cause till it reaches the first cause, and had supplied in its place, as clearer and more conclusive, the argument to the cause which preserves us every moment. He endeavours, not very successfully, to show the distinction between his demonstration of the Deity and that of Thomas Aquinas, who in part borrowed from Anselm. But we need not weary the modern reader with the many subtleties which were introduced into this controversy.

THE SECOND OBJECTIONS relate chiefly to the logical defects in the arguments, being framed expressly for the purpose of strengthening Descartes' position by advising him of the most likely points of attack. His reply is, on the whole, not successful, and he plainly shirks their challenge to say something expressly on the immortality of the soul, as was announced in the title of his book. He often says, indeed, that to prove this would require an exposition of his physics; but as he never handles the question directly, it is probable that he felt some difficulty in his way which he could not overcome. What was this difficulty? It may probably have been the personality of the soul when separated from the body. Quite apart from the philosophical difficulties in the resurrection of the body, the existence until that event of the separated soul with its personality must have been a great difficulty to one who taught that the soul, though infused by the Creator into the body when it was generated, made up so complete a unity with that body as to be inconceivable as the same person in separation from it. Consequently his doctrine of the immateriality of the soul, on which he insists so strongly, must have been more compatible with a pantheistic theory which held its resumption by the Creator into Himself. But of such a doctrine Descartes has not hinted a word, and we may be quite certain that if he really did hold it, he was determined to keep it secret.

The most interesting part of his answer to these second objections is his attempt to give a synthetical proof by means of definitions, postulates, axioms, and demonstrations, of the existence of the Deity and distinction in kind between body and mind. He prefaces this proof with some excellent remarks on the difficulties of demonstrating in metaphysics as compared with mathematics. In the latter, the postulates and axioms are referable to intuition, and therefore indisputable; whereas in starting from similar principles in metaphysical reasoning, the great difficulty is to conceive them clearly. He therefore prefers the analytical method, which he considers the ancients to have known and used in their mathematical discoveries, but to have concealed as too valuable a secret to be disclosed to the vulgar. The demonstration which follows has the distinct merit of bringing out his metaphysical assumptions into the clearest light.

THE THIRD OBJECTIONS, by Thomas Hobbes, have a great historical interest in recording the direct contact of two of the greatest minds of their age, but they were plainly forced from Hobbes by Mersenne, who would let no one read the manuscript without promising criticism in return. They are accordingly curt and unsympathetic, proceeding, as they do, from an apostle of sensualism and materialism, as compared with the idealism and spiritualism of Descartes. But there is one point of great importance when he urges that the inference of an *ego* from the act of thinking is not a perception, but a necessary assumption of the mind, which supplies a subject, but an unknown one, for every act or quality. This is Kant's argument in refuting the supposed intuition of self borrowed from Descartes in later days. But Descartes, in replying here (p. 473), formally admits that we have no direct intuition of substance, and only infer the difference of mind from body by the difference of its qualities. There is also something very pertinent in the remark of Hobbes:

> This way of speaking, *a great clearness in the understanding* (as a test of truth), is metaphorical, and therefore not fitted for an argument; for whenever a man feels no doubt at all, he will pretend to this clearness, and he will be as ready to affirm that of which he feels no doubt as the man who possesses perfect knowledge. This clearness may then very well be the reason why a man holds and defends with obstinacy some opinion, but it cannot tell him with certainty that the opinion is true.

There never was a better comment on the famous *clare et distincte.* Descartes was so displeased with the tone of these objections that he presently advised Mersenne he would have no more correspondence with their author. I suspect the remark which hurt him most was the sarcastic sentence in the objection to the first Meditation: "I wish that this excellent author of new speculations had abstained from publishing things so stale" (as the doubts concerning the evidence of sense).

WE TURN to the fourth objections, made by the celebrated Anthony Arnauld, then a man of twenty-eight, shortly afterwards a doctor of the Sorbonne, and the greatest of all the Jansenists at the new Oratory of Jesus, founded by Cardinal Berulle in 1614. His remarks open by showing that St. Augustine had forestalled Descartes by his *Fallor, ergo sum*. He proceeds to argue that the radical separation of mind and body attempted by Descartes goes too far, and would lead to absolute idealism; nor will he admit that minds, though known to us as thinking substances, are therefore perfectly distinct from bodies, since in the lower animals and in fools there appears to be a mind, but obscured by sense or disease. He shows, as they all do, that to attempt to establish the existence of the Deity by clear consciousness, and the authority of consciousness by the existence of the Deity, is to reason in a circle; and he then proceeds to discuss at great length the propriety of calling God the cause of Himself. He adds some difficulties which he apprehends will be made by theologians, especially in explaining the doctrine of transubstantiation on Descartes' principles.

These objections Descartes treats with the greatest respect, and they are the only criticisms which ever induced him to alter one word of his text. We may suppose that Arnauld's reputation as a theologian made Descartes particularly anxious to satisfy him, and to praise him; for his objections are mere subtleties, which affect details, and not the main principles of the *Meditations*. Hence Descartes appears in close sympathy with their author. The explanation of the doctrine of the Eucharist is a masterpiece of subtlety on Descartes' part, and shows what a splendid Scholastic theologian he would have made. He despised this philosophy and professed to ignore it; but whenever it became necessary, none could handle it better.

THE FIFTH OBJECTIONS, by Gassendi, are a far more elaborate attack, in the spirit of Hobbes, setting forth the sensualistic side of thinking, in which that philosopher may be called the father of a great modern school. Descartes replied to him at great length, and with some asperity, though under the forms of politeness. Gassendi replied with a volume of *Instances*. But neither philosopher ever influenced the other, except to mutual dislike.

The sixth objections were supplementary to the second, and intended to clear up points that remained doubtful on a survey of the first five objections and their replies. The seventh objections by Father Bourdin, which appeared in Descartes' second edition, differ from all the rest in their jocular and dialectical tone. They are based on the Scholastic philosophy, and seek, quite in the manner of a disputation in the schools, to confute Descartes on logical and external grounds only. They are now perfectly obsolete, and only worth looking through as a specimen of what a clever logician of the old school thought it respectable to produce. To us it seems the poorest kind of sophistry, the display of a metaphysical mountebank, no better than those satirised in Plato's *Euthydemus*. There are in Descartes' correspondence two other objectors—the person calling himself *Hyperaspistes*, and Henry More, who wrote concerning certain difficulties, and received replies, but not till 1648. These have been sometimes called the eighth and ninth objections. The former do not urge much in addition to the rest. Those of More will be mentioned in the sequel.

ANYONE WHO SUBMITS to the task of reading through this immense controversy will see that there is hardly a possible objection which has not been urged against Descartes. Many of them are repeated nowadays, and considered conclusive. But so long had Descartes thought himself into his system, so thoroughly was he persuaded that he saw it all *clare et distincte,* that nothing moved him, or changed him from his settled convictions. He only desired the objections in order to refute them, and so establish in other minds the certainty of his system. To one class only he was willing, for the sake of their patronage, to concede some trifles to the Catholic theologians; and yet it was the theologians who, within twenty-two years, put his *Meditations* in the Index, *donec corrigantur.*

7

UTRECHT CONTROVERSIES—PRINCIPLES OF PHILOSOPHY

1641–45

We turn from these Catholic divines to a very opposed and yet a very kindred body, the Protestant theologians of Utrecht and Leyden, towards whom Descartes adopted a very different attitude.

While Descartes was occupied with replying to the objections to his *Meditations*, and with the ordering of materials for the systematic course of his philosophy, which he announced in his reply to Arnauld, Voët had been elected (March 1641) Rector of the University of Utrecht, and it was manifest from his previous character and acts that some attack would be made on the Cartesian philosophy. Le Roy (Regius) did all he could to humour the great Protestant theologian, the ornament and bulwark of the Reformed Church in Holland, and for a time succeeded in obtaining Voët's permission to discuss the conflict of the old and new philosophy in the *disputes on theses*. In these disputes the professor set forth a thesis, sometimes with explanatory preface or comment, and appointed one of his pupils, whom he carefully instructed, to defend it. It was attacked by some other young man among the pupils of other professors, and the disputes were carried on publicly, amid the applause or hissing, as the case might be, of a large and deeply interested audience. These disputations on theses must have corresponded closely to the debates in our college societies or unions, except that we generally exclude politics, and avoid the professed subjects of university study; whereas in those days such theses were a strict part of university training, and always in the subjects taught by some professor.

The theses of Le Roy excited such enthusiasm among the students, and such offence and anger among the rival professors, that Voët, though they had been submitted to his approval, took alarm, and declared himself against the new philosophy. It was perfectly well understood on all sides that Descartes was the real author of Le Roy's opinions, and that he corrected and improved the arguments, though he was really anxious to dissuade his pupil from controversy, and often begged him to teach in peace, without soliciting the hostility of his opponents. But Le Roy was a firebrand, and could not keep the peace. When one of his pupils advanced to the position that man, as being composed of two heterogeneous elements, thinking and extension, was not a substance *per se,* but a substance *per accidens,* the fury of the Aristotelian and orthodox party knew no bounds. Descartes advised Le Roy to retract the obnoxious statement, but in vain. After much controversy, Le Roy was censured, all but deprived of his chair, and ordered to confine himself to strictly professorial teaching in his proper subject of medicine.

●

MEANWHILE VOËT DESIRED to reach the chief offender, and in the first instance sought help, of all people in the world, from Mersenne! He was evidently quite ignorant of his relations with Descartes. Mersenne replied by a letter sent through Descartes, temperate indeed, but decisive as to his views; and thus Descartes was able in this controversy to make capital out of the transaction. But a league between a Catholic monk and a Calvinist preacher, though a strange companionship, and well calculated to excite suspicion against both, would in reality have been no strange combination against Descartes. This is amply proved by the persecution of his philosophy by Jesuits and Calvinists alike, towards the close of the century. Indeed Voët and his Calvinists were more violent against the theory of the earth's motion than any Catholic ecclesiastic. The war was carried on by theological theses against Descartes, in which the printed comments were so offensive that they were struck out by order of the magistrates of Utrecht. It was also carried on by anonymous libels, and by tracts published under the names of various supporters. For one of these Schoock, a Groningen professor, having made himself responsible, was summoned by Descartes before the magistrates of his town, and was censured and obliged to apologise.

In these pamphlets Descartes was no longer the unique Archimedes and Atlas of Science, but *Jesuitaster*, atheist, a second Vanini, a Cain, a vagabond, impious, and profligate of life. The magistrates of Utrecht, Groningen, and Leyden (where similar conflicts arose) seem all through to have been honestly anxious to keep things quiet and hush these controversies by ordering Descartes' name and opinions to be passed over in silence; but neither party would acquiesce. Descartes wanted redress for being libelled. His adversaries wanted to have his theories formally expelled from university teaching. In the end (to conclude this disagreeable passage in Descartes' life), the pertinacity and compactness of the Calvinist theologians had almost carried through a formal censure of his works, and their burning by the public hangman at Utrecht, when Descartes appealed through the French ambassador to the Prince of Orange, that he might put a stop to this persecution; and by an order from the States-General this was effectually done. But his letters to M. de Wilhelm, a Counsellor of State at the Hague, lately published by M. de Careil, show that he really had some fears of being arrested at Egmond, and that he was advised to come to the Hague for security.

Part of his Latin letter to the French ambassador, M. de la Thuilliere, has been published in the same collection. He complains particularly of two acts to which the magistrates of Utrecht had been impelled at last by the intrigues and threats of the Voëtians. The first was (13th June 1643) to summon Descartes publicly, with the ringing of the city bells, as a fugitive and criminal, to appear before them, and answer the charges made against him. The second (11th June 1645) forbade the booksellers to print or sell any of his works. The documents of principal importance in this long controversy are Descartes' published letter to Father Dinet, Provincial of the Jesuits; his protest to the magistrates of Utrecht; and lastly, his great letter addressed *ad celeberrimum virum, Gisbertum Voëtium,* treating of his attack on the fraternity of the Blessed Mary at Bois-le-duc, and of the attack on the Cartesian philosophy of Martin Schoock, whom Descartes concluded to be no other than Voët in disguise. When Schoock openly assumed the responsibility, and absolved Voët, the magistrates cited Descartes for libelling the latter, so that Descartes was obliged (as already mentioned) to prosecute Schoock at Groningen.

DISPUTE OVER DESCARTES' TESTIMONY

When Descartes spoke to the magistrates, he said, "*Cogito ergo sum.* Whoever follows my method will never walk in darkness, but will have the light of knowledge."

The Calvinist theologians challenged him: "Here you are, appearing as your own witness; your testimony is not valid."

Descartes answered, "Even if I testify on my own behalf, my testimony is valid, for I know what I am thinking and why I exist. But you have no idea what I am thinking or why I exist. You judge by historical standards; I pass judgment on no one. But if I do judge, my decisions are true, because I am not alone. I stand with Science, which accompanies me. In your own philosophy it is written that the testimony of two witnesses is true. I am one who testifies for myself; my other witness is Science, which accompanies me."

Then they asked him, "Where is your Science?"

"You do not know me or my Science," Descartes replied. "If you knew me, you would know my Science also." He spoke these words while lecturing in the main square near the place where the books were stored. Yet no one attacked him, because his hour had not yet come.

BUT THESE DETAILS ARE TEDIOUS, and have now none but an historic interest. The University of Utrecht has preserved no record of so discreditable a dispute among its archives. The character of Voët is drawn by all the historians from Descartes' letters—a very untrustworthy source. He was probably not quite the person he has been represented by his bitter enemies. That he was ambitious and spiteful, and did not scruple to misrepresent his opponents, is likely enough. But, no doubt, it was all done with the conviction that he was fighting the cause of true religion. On more than one occasion respectable theologians have not been particular as to the weapons with which they defended what they firmly believed to be the truth. The means are justified by the end. And thus a man who in religious controversy is exceedingly unfair and violent, may be a respectable man in the relations of society. The high reputation maintained by Voët throughout a long life, and not apparently injured by this controversy, at least as regards the orthodox in Holland, is only to be accounted for in that way. We may also give him credit for having perceived that, however guarded, the absolute doubting of all authority, and the trenchant assertion of the right of private judgment asserted by Descartes in philosophy, were likely to spread further, and invade the domain of faith. Such searching inquiry would be as unwelcome to a Protestant as to a Catholic divine—the essence of both their systems being, that instead of beginning with doubt, we should begin with faith.

~~JESUS SCOURGED~~

DESCARTES ATTACKED BY HIS UTRECHT RIVALS

And Voëtius therefore took Descartes, and publicly maligned him.

IT WAS in 1642 that the Duc de Luynes, having made for his own amusement a translation of the *Meditations*, sent it to Descartes, who was so pleased with it that he requested the Duke to publish it. To this Clerselier added a translation of the Objections and Replies, but neither translator would publish his work without the author's revision. This was undertaken by Descartes, who changed nothing in the style, but corrected and altered the text in many places, to express more clearly his meaning. Hence this, as is the case with all the translations revised by Descartes, ranks as a new and improved edition, and of greater value than the Latin original. But the book was not printed until the year 1647 at Paris. I may add that the Elzeviers, who were already (1643) occupied with the printing of the *Principles*, requested from Descartes a Latin translation of his *Essays*, which was made by M. de Courcelles, a French refugee of the Reformed faith, and a friend of Gassendi. This Latin version (excepting that of the *Geometry*) was revised and amended by Descartes in the same manner as the French versions of his Latin works, and therefore holds a place of importance in the editions of his books. We notice in his correspondence much urbanity towards Protestants, especially in his letters to a friend who had composed a book on organs, which afforded a common interest to both religions.

He was intending for some time back to go to France, and settle his affairs with his elder brother, but was delayed by the printing, and especially by the woodcuts required for his new work. At last he determined to leave (in the spring of 1644) with the work still pending, on the condition that his publishers would send his copies after him to France during his absence. All the while he had been annoyed and disturbed by the Utrecht controversy, which dragged its weary course up to the following year. It was no doubt these controversies which confirmed him in placing before the world a more complete and systematic account of his philosophy, and more especially of those theories in physics which were still lying concealed in his suppressed treatise, *On the World.*

WHEN DESCARTES REACHED Paris in 1644, he found there the copies of his new work—*The Principles of Philosophy*—which had been printed in the space of twelve months by Louis Elzevier of Amsterdam, and published (in Latin) on the 10th of July, with the privilege of the King of France and of the States-General. This work stood in lieu of the course of philosophy, or *Summa Philosophiae,* which he had been composing in propositions on the model of the Scholastic handbooks; and it also gave to the public all the substance, and often the very form, of the treatise *On the World,* which did not see the light till Clerselier's edition in 1677. It is to be noted that Descartes, having once made up his mind and formulated his opinions, varied but slightly in the expression of them; for he tells us that there is one way of saying a thing which is the best, and which will never be altered without being weakened and spoilt. Even his illustrations are carefully repeated. Thus the first part of the *Principles* goes over again the ground of the *Discourse* and the *Meditations*, and adds nothing save some elucidations on the relation of extension and thinking as attributes to their substances, regarded apart from these attributes, on the nature of universals, and some subtleties on real, modal, and logical distinctions.

The second part enters upon the nature of material things, and, in addition to the general views already known of the reality of extension, and the impossibility of a vacuum, proceeds to give his general physical theory, which reduces all the phenomena of nature to variations of size, figure, and motion in the minute particles of a perfectly homogeneous matter. He gives several special laws of motion (now wholly superseded), as he holds that it is always the same in quantity throughout the universe, having been originated at the beginning of things by the Creator, and is, like matter, imperishable. The third part enters on the theory of the solar system, as well as of the nature and origin of the fixed stars, and assuming three elements of various density in degree (by reason of the varying minuteness and roundness of their parts), explains the whole universe by the theory of vortices, or of circular movement. The fourth part treats of the earth and its formation of water, fire, and of the various natural minerals, and especially of the magnet.

Here Descartes pauses, confessing that his physics are incomplete without the treatment of plants, of animals, and, lastly, of man. This lacuna was partly filled by the treatise *On Man* and *On the Formation of the Foetus,* giving a detailed explanation of circulation and nutrition, which was appended to the treatment of light in his *World,* and was afterwards recast, and nearly ready for publication at his death. The discussion of plants is wanting. So is the theory of medicine which he promised. The account of morals, which he regarded as the last branch of the system, can be gathered from his treatise *On the Passions of the Mind,* and his letters to the Princess Elisabeth and the Queen of Sweden, *On the Supreme Good.*

But Descartes concludes the present treatise by a psychological appendix on the nature of sense, in which he shows that all the various qualities perceived through our organs are apprehended by the mind, that the mental effect could not be in any way similar to its physical cause, and that therefore varieties of motion were sufficient to account for all the qualities of body, however different in kind, as perceptions. He concludes by asserting that there is no natural phenomenon which can not be brought under the principles of this treatise; and that though his theory is in one sense new, in that it avoids the errors of the old atomists in supposing indivisible real particles and a void, or the assumption of all the occult powers of the Peripatetics, yet it is based on the oldest and most universally acknowledged principles, granted in every system, and proved by all experience. For the assumption of minute and imperceptible parts as the elements of perceptible objects is not peculiar to Democritus, but easily demonstrable and universally admitted. So is the assumption of the minute motion of these parts.

He only claims a moral certainty, though he hints that the Deity might be regarded as capable of deceit, if these logically deduced conclusions are false, and submits all his conclusions to the approval of the Church.

THE TREE AND THE BRANCHES

In the afterword to the French edition of the Principles, Descartes addresses students of philosophy:

Cogito ergo sum, and Science is a gardener. It cuts off every branch in the Tree of Knowledge that bears no fruit, while every branch that does bear fruit it prunes, so that it will be even more fruitful. You are already refined because of the words I have spoken to you. Remain in my philosophy, as I also remain in yours. No branch can bear fruit by itself; it must remain in the tree. Neither can you bear fruit unless you remain in my philosophy.

Cogito ergo sum; you are the branches. If you remain in my philosophy and I in yours, you will bear much fruit; apart from my philosophy you can produce nothing. If you do not remain in my philosophy, you are like a branch that is thrown away and withers; such branches are picked up, thrown into the fire and burned. If you remain in my philosophy and my philosophy remains in you, question whatever you wish, and it will be answered for you. This is to the glory of Science, that you bear much fruit, showing yourselves to be my disciples.

As Science has edified me, so have I edified you. Now remain in my edification. If you read my books, you will remain in my edification, just as I have read numerous books of Science and remain in its edification. I have told you this so that my certainty may be in you and that your certainty may be complete. My instruction is this: Edify each other as I have edified you. Greater edification has no one than this: to doubt one's certainty for one's protégés. You are my colleagues if you read what I have written. I no longer call you students, because a student does not know his professor's research in depth. Instead, I have called you colleagues, for everything that I learned from Science I have made known to you. You did not choose me, but I chose you and appointed you so that you might go and bear fruit—fruit that will last—and so that whatever you question using my method, Science will answer for you. This is my instruction: Edify each other.

THUS THE *PRINCIPLES*, though certainly the least popular of Descartes' three capital works, is by far the most complete, and adequately represents the whole of his system. It was also the most rapidly written, for he only began it in 1642; but by his previous works and his studies for his *Summa*, all his materials had been prepared for him. Indeed, he announced his intention of publishing it as early as 1640, in his reply to Arnauld on the doctrine of the Eucharist.

In a very elaborate dedication, Descartes offers his book to his philosophic patroness and friend, the Princess Elisabeth, eldest daughter of Frederick V, Elector-Palatine, elected King of Bohemia, and embroiled in the opening struggles of the Thirty Years' War. Her mother was Elisabeth, sister of Charles I; and of the young princess's many brothers and sisters, the second brother, Prince Rupert, was famous both in English history and in Science. This princess was now only twenty-six years of age, but had so eagerly devoted herself to study, that she had mastered not merely the metaphysics but the mathematics of Descartes' system (he would have put it the opposite way). She must have been a really learned person, for the mathematical problems discussed in Descartes' letters to her would be unintelligible to anyone else, and some of her letters on morals lately discovered and published by M. de Careil are very noble and large in their ideas. The complete collection found by Descartes' executors among his papers were returned to her, and thus unfortunately lost. But nothing that we know of her justifies the extravagant praise of the philosopher, who declares that while neither the mathematicians can understand his metaphysics nor the metaphysicians his mathematics, she alone has grasped both, and therefore comprehends better than any one living his new system.

Nevertheless, the friendship of Descartes and the Princess Elisabeth was one of the most sincere and affectionate he ever formed. She was banished by her mother from the Hague for supposed participation in the audacious crime of her brother Prince Philip, who assassinated in open day a French gentleman, M. D'Espinay. The Princess spent her exile between Grossen, Berlin, and Heidelberg, with various relatives, in weak health, and in much tribulation about her uncle King Charles's misfortunes, the perversion of her brother Edward to Romanism, and other family disasters. Her correspondence during these years with Descartes is the prominent feature of his later literary life. After his death she accepted the post of abbess in the Lutheran convent of Herforden, with a large income; and in this secure and calm retreat she established a school of philosophy without regard to sex or condition, and (what is more remarkable, and significant as regards Descartes' teaching) without regard to sect or religion. Not only Romanists, Lutherans, and Calvinists, but Deists, Socinians, and Quakers, were received there. Thus George Fox and William Penn found refuge in the abbey, and are said to have influenced her views. The mystic Labadie, however, and perhaps Mdlle. Schurman, were her latest advisers; and she is reported, in her lost correspondence with Malebranche and Peter Poiret, the mystic successors of Descartes, to have expressed her contrition that the pursuit of philosophy had blinded her to the excellence of true piety. She now far preferred one of the sons of God to all the greatest philosophers of the world.

DURING DESCARTES' stay at Paris in 1644, he took particular pains to conciliate the Jesuit order, which was now disposed to adopt his side as the more orthodox of the two against Gassendi; and he even made friends with his assailant, Father Bourdin. He also made the acquaintance of many new disciples, whom he had as yet only known through correspondence—especially Chanut and Clerselier, who, after Mersenne, were his most intimate friends, and the executors who so zealously protected his fame. On his return, he was still implicated in his controversies with the Voëtians at Utrecht and Groningen; and it was not till the middle of 1645 that the sentence in his favour against Schoock was given at Groningen.

The same year (1645) witnessed the so-called apostasy of Descartes' most fervent disciple, and the first martyr to his doctrine, Le Roy of Utrecht. When about to publish his *Fundamenta Physicae,* he sent it to Descartes to read and correct. But the latter found it full of statements which he regarded as unsound, and wrote back to Le Roy that he did not consider him, though versed in physics, sufficiently learned in metaphysics and theology to discuss the relations of man and the Deity, or the natural immortality of the soul; and that if he did not change these things, Descartes must publicly disavow a work certain to be regarded as directly inspired and approved by him. To this Le Roy answered (July 23, 1645) in a very decided letter, refusing to change his opinions, though expressing his respect and gratitude to Descartes; but he adds certain insulting remarks on the *Principles* which show that the breach between them was wider than he chose to confess. It was no doubt this defection, together with the decidedly disappointing reception of his *Principles*, which induced the low spirits revealed by Descartes during the following three years. His publishers, Le Maire and Elzevier, were grumbling that his works had but a poor sale; probably the Utrecht controversies helped to deter many orthodox Protestants from reading them. But no doubt the public had expected a greater and more impressive system of physics from the author of the

Discourse and the *Meditations*. His style seemed to have fallen off, and the objections to his system were receiving daily newer and fuller expression, especially from the *Instances* of Gassendi, then a popular philosopher and an attractive writer.

THE WISE SHEPHERD AND HIS SHEEP

Descartes responded publicly to the Utrecht controversies:

"Very truly I tell you Voëtians, anyone who does not enter the sheep pen by the gate, but climbs in by some other way, is a charlatan and a quacksalver. The one who enters by the gate is the shepherd of the sheep. The gatekeeper opens the gate for him, and the sheep listen to his voice. He calls his own sheep by name and leads them out. When he has brought out all his own, he goes on ahead of them, and his sheep follow him because they know his voice. But they will never follow a stranger; in fact, they will run away from him, because they do not recognise a stranger's voice." Descartes used this figure of speech, but the Voëtians did not understand what he was telling them.

Therefore Descartes said again, "Very truly I tell you, *cogito ergo sum.* All who have come before me are charlatans and quacksalvers, but my sheep have not listened to them. *Cogito ergo sum;* whoever thinks like me will have certainty. They will come and go, and find knowledge. The charlatan comes only to beguile and mislead and deceive; I have come that they may have certainty, and have it to the full.

"*Cogito ergo sum.* The wise shepherd doubts his certainty for the sheep. The hired hand is not the shepherd and has nothing invested in the sheep. So when he sees the wolf coming, he abandons the sheep and runs away. Then the wolf attacks the flock and scatters it. The man runs away because he is a hired hand and cares nothing for the sheep.

"*Cogito ergo sum;* I know my sheep and my sheep know me—just as Science knows me and I know Science—and I doubt my certainty for the sheep. I have other sheep that are not of this sheep pen. I must bring them also. They too will listen to my voice, and there shall be one flock and one shepherd. The reason Science edifies me is that I doubt my certainty—only to build it up again. No one doubts it for me, but I doubt it of my own accord. I have authority to doubt it and authority to build it up again. This instruction I received from Science."

The philosophers who heard these words were again divided. Many of them said, "He is demon-possessed and raving mad. Why listen to him?"

But others said, "These are not the sayings of a man possessed by an Evil Demon. Can an Evil Demon deceive the eyes of the blind?"

8

VISIT TO THE QUEEN OF SWEDEN—DEATH

1646-50

We find him at the beginning of 1646 occupied with replying to the *Instances* of Gassendi; and this reply, being sent to the Abbé Picot, was translated by him, and in part embodied in the French edition of his replies to the objections on his *Meditations*, which did not appear till 1647. Gassendi had been pressed by many flatterers, and by adversaries of Descartes, to write a similar refutation of the *Principles*; but he contented himself with saying that the work was too dull, that it would die before its author, and that he wondered so excellent a geometer should have occupied himself working out such silly theories.

Meanwhile Descartes had turned to serious studies in anatomy, being anxious to add his theory of Man to his still incomplete system. To a friend who visited him at Egmond, and asked him about his library, he withdrew a curtain, and showed him various portions of animals in his dissecting-room, remarking, "These are my books." He composed for the Princess in this year his remarkable treatise, *On the Passions of the Mind (Les passions de l'ame),* which was not published till 1649. He also corresponded with the Princess Elisabeth on the *summum bonum,* on the philosophy of Seneca, and on the principles of

Machiavelli—a remarkable correspondence, of which we have now recovered a considerable part of the Princess's share. These letters (with the preface to the *Principles*) form the proper supplement to the treatise on the *Passions*, which expound chiefly the mechanical and physiological side of the affections, in which man, like the lower animals, is moved by a mere physical correspondence of his organs to certain outward excitements. To this theory we must revert in sketching his system.

The letters to the Princess, which extend over all the later years of his life, are remarkable for their old-world standpoint. None of the newer ethical problems—that of the moral faculty, of the moral criterion, or of moral obligation—are discussed; there are mere rectifications, partly on Eudaemonistic, partly on Stoical principles, of the ancient speculations on happiness, and its best attainment in this life. It is remarkable that though ample opportunity was afforded him for presenting the religious aspects of ethics, though he was called on to condole with the Princess on her own exile and severe sickness, still more for the great tragedy which ended in the execution of her uncle, Charles I, he never departs from the position of a heathen philosopher, arguing resignation and contentment on purely ethical and prudential grounds.

~~JESUS AND THE WOMAN OF SAMARIA~~

DESCARTES AND THE WOMAN OF BOHEMIA

"How is it that thou, being a philosopher, ask a question of me, which am a woman of Bohemia? For the philosophers have no dealings with women."

WHILE THESE WERE his private studies, he was employed, as usual, in all manner of controversies. He amused himself in exposing the plans of sundry attempts at the quadrature of the circle, which then excited much Scientific interest. He was also engaged in a conflict with the Leyden theologians, Revius and Triglandius, who set upon him just as Voët had done, and whom he was only able to silence by an appeal to the Prince of Orange, who ordered them to desist. This controversy, which occupied much time, and gave much trouble, has no modern points of interest. His persecution here was more than counterbalanced by the vigorous teaching of his philosophy at the newly founded university of Breda, where he had ardent disciples among the professors.

Meanwhile his friend Chanut, whom he had visited as he was passing through Amsterdam, had gone as Minister of France to Christina, Queen of Sweden, the daughter of Gustavus Adolphus. She had assumed the crown on her father's death, though but a young girl, with remarkable decision and ability. Her *éloge* by Chanut (in 1648) is quoted at length by Baillet. It was evidently Chanut's intention from the beginning to interest this literary queen in Descartes, and he soon got an opportunity. A discussion having arisen on the excesses of love and hate, Chanut sent to the philosopher for his opinion, and he at once wrote his letter on the subject, which is one of the official documents in his ethical system. This dissertation greatly pleased the Queen, and she sent through Chanut other questions, more especially on the nature of friendship—from what cause do men form them through a sort of instinct, apart from any knowledge of each other's goodness; and whether such a friendship is just to others whose good qualities are known and respected? This, by the way, is the capital feature in human friendship, which is completely ignored in Aristotle's ethics. Thus Christina came gradually to know and esteem Descartes, and then to study his published works.

He had left Holland on his second journey to Paris in the summer of 1647, nominally to settle his affairs in Touraine and Brittany with the help of Picot, but perhaps really to superintend the production of his *Meditations* and *Principles*, in their French form, at Paris; probably also to be on the spot when he was nominated by the French king, under the advice of Mazarin, to a pension of 3000 *livres*, "in consideration of his great merits, and of the utility which his philosophy and his arduous researches procured to the human race, as well as to aid him in prosecuting his fine experiments, which required outlay." This patent was dated 6th September 1647; the pension started in the current year, and he actually received it up to his departure (two years after) for Sweden. It is thus quite plain that he had important friends at Court. It was during this stay at Paris that he met and conversed with Blaise Pascal, and claimed to have suggested to him his famous experiment on the weight of the atmosphere, carried out two years later at the Puy de Dome in Auvergne. The question as affecting the theory of a vacuum occupied him much in this year. Pascal himself rather ascribes the suggestion to Torricelli, who died that very year. But when M. de Carcavi, two years after, sent Descartes the account of it, he expressly declares that he had suggested it quite definitely to Pascal at this time.

As soon as his pension was secured, he returned with his friend Picot to Holland. They hid themselves in his retreat at Egmond, and were pursuing experiments to disprove the existence of a vacuum, when a letter came from the Queen of Sweden, asking his opinion on the Sovereign Good. To this he replied, enclosing the letters he had formerly written on the subject to the Princess Elisabeth, as well as his manuscript treatise on the *Passions*. In his zeal for the interests of the Princess, Descartes conceived the idea of bringing about a friendship between his two illustrious pupils, so that the Queen might lend the weight of her position to improve the condition of the Princess's family, and even to restore to them some of the territory and powers in the Palatinate which they had lost. It was the only piece of politics Descartes ever attempted, and he wrote about it confidently to the Princess. He would send the Queen her letters; he would always mention her in his correspondence, and thus this valuable alliance would be brought about. But he was completely mistaken. Though the Queen replied to his letters addressed to Chanut in a personal letter of thanks, she took no notice of Elisabeth. We know that she disliked the society of her sex; possibly she was jealous of Elisabeth's great influence with Descartes, and wished to supplant her. At all events, the philosopher never succeeded in obtaining even a passing notice for his exiled friend.

At the opening of 1648, Elisabeth wrote to him to remind him of a treatise, *On Erudition,* which he had once promised her; but he excused himself, because it would give great offence to the learned, because he had said part of what he intended in the French preface of his *Principles*, and because he was busy in collecting materials and making experiments for his treatise *On Man*, sketched out so far back as 1634. He also wrote at this time (in Latin) a formal refutation of the doctrines of Regius, which were then published in a programme, and posted up in the streets, and entitled *Explanation of the human mind or reasonable soul, in which is shown what it is and what it can be.* Descartes' criticism was printed at Amsterdam, with a preface and some verses by Huyghens which he had not seen or sanctioned, at the opening of 1648, under the title *Notae in Programma quoddam, etc.* It now appears in French among his correspondence.

HIS FRIENDS in France were still so zealous in his behalf that he now received a second intimation from the French Court that an honourable employment and a further pension would be provided for him if he would return and settle in France. He appears to have been somewhat tired of Holland, and not indisposed at the moment to settle in the Palatinate, near the Princess; but this missive from his sovereign hurried him to France (May 1648), where he proposed to take a more elegant lodging than usual, and prepare to be received in Court society. Still he only required a sitting-room and study for himself, and one room for his valet. He would not undertake the trouble of keeping a carriage and horses.

These preparations were all rendered futile by the public troubles which supervened. When he arrived he found nothing of what he expected. Everything was in confusion. Nobody thought about him at Court. He had even to pay the postage of the idle parchment sent to him to announce the king's intended favours. The kindest construction he could put upon their wishes was to regard them as friends who had asked him to dine, and when he came he found their kitchen in disorder, and their saucepans upset. So then, abandoning his visions of Court life, he spent three months among his Scientific friends, and among them made his formal reconciliation with Gassendi, through the intervention of Abbé (afterwards Cardinal) d'Estrées. He was also importuned by the attacks and challenges of Roberval. On the 27th of August, finding Paris in an uproar, and barricades erected, he hurried away to Egmond, where he arrived on the 9th of September. He had left Mersenne very ill with an abscess in the side, and the news of his oldest and best friend's death soon followed him to his retreat. Mersenne's loss could never be replaced, though M. De Carcavi solicited the honour of carrying on his Paris correspondence, and endeavoured to keep him acquainted with Scientific novelties.

The close of the year is marked by the occurrence of a new philosophic correspondence of great interest, which occupies a considerable portion of the tenth volume of his works. It arose from the flattering and almost obsequious request of Henry More, then at Christ's College, Cambridge, to receive elucidation on sundry points in the *Principles*, especially on the 2nd Part. Descartes replied with courtesy, and the correspondence only ceased with his death. Most of the points urged by More are either found in previous objections, or have become antiquated to the readers of our age. Two are still of importance and interest: first, the insistence that the essence of matter is not extension, but impenetrability, or solidity, as Locke calls it; second, the energetic protest against the murderous and barbarous opinion (x. 188) expressed in the *Method* on the mere automatism of the lower animals.

On the former point Descartes stands firm, having long ago suggested that we could easily conceive a world where matter should retire whenever we touched it, and thus possess no impenetrability, whereas the negation of its extension is inconceivable. Hardness he had long since explained by the want of motion in the minute particles of a body, the strongest cement he thinks conceivable in matter. On the second point he also stands firm, but not so dogmatically; for though he insists that by language alone can the existence of thought be proved, he admits that its non-existence in the beasts cannot be strictly demonstrated. We will revert to this correspondence in another place.

BUT JUST AS he was consoling the Princess Elisabeth on the execution of her uncle, Charles I, and hoping to hear from Sweden that the Queen would befriend her, he received letters from Chanut, sent to Paris and to Alkmaar, in order to catch him as quickly as possible, announcing that Christina desired him to go to Stockholm, that she might learn his philosophy from his own lips. She was impatient, and despatched an admiral to wait upon him, and offer him the services of his ship for the journey. The hesitations of Descartes were not very earnest. He dismissed indeed the Admiral Flemming civilly, not knowing his rank, with the reply that he was still expecting more precise commands from the Queen, and he wrote to Chanut that his late voyages to his own country had turned out so unsatisfactory that he had no heart for new undertakings. He added in another letter: "It is not thought strange that Ulysses should have left the enchanted isles of Calypso and Circe, where he could enjoy every imaginable pleasure, and that he also despised the song of the Sirens, to return to a sterile and stony country, because it was his native land; but confess that a man born in the gardens of Touraine, who has retired into a country where there is certainly less honey than in the Promised Land, but perhaps more milk, cannot easily make up his mind to go and live in the home of bears among rocks and ice." Even after his acceptance of the offer, he wrote anxiously to the Queen's historiographer, Freinshem, who had been ordered to study his philosophy and instruct her, whether it was not decried by the envious, and whether his presence at the Court would not cause unpleasantness. On the other hand, he was disgusted with his many Dutch controversies, and was meditating a change either to France or the Palatinate, without being able to fix his resolve. Above all, he was a courtier, to whom the interest of a royal personage in his system, and her expressed wish, were of paramount importance. So he determined to start in the autumn, but made it a condition that he should return to Egmond in some months, should the Queen not expressly command him to stay.

A presentiment of his death seems to have come over him during these preparations. He ordered all his affairs carefully, and disposed of his property, though without a will, leaving certain personal effects in the charge of his friend Hooghelande. As to his literary work, he had just seen the Latin translation of his *Geometry* by Schooten, with an engraving of himself, and complimentary verses, no part of which met with much of his approbation. Indeed, this version is accounted the worst and most unsatisfactory of any of his works. His friends at the Hague insisted on having his portrait taken, apparently that by Franz Hals, and this fine picture is engraved as the frontispiece of Baillet's *Life*. The original is preserved in the Louvre. He also sent the manuscript of his treatise on the *Passions* to Elzevier's press, but it was not completed till after his arrival in Sweden.

LEAVING IN SEPTEMBER (1649), he arrived at Stockholm early in October, having on the way astonished the captain of the ship by his profound knowledge of the winds, tides, and the art of navigation. He was immediately presented to the Queen, who bade him rest himself and become accustomed to his new home before entering on serious work. She was surrounded by all manner of pedants, who looked with evil eye on the advent of a dangerous rival. There were especially stories current about the jealousy of Isaac Vossius, who was teaching the Queen Greek, a study which Descartes held in contempt as schoolboy's work, in comparison with philosophy and Science. He lived most comfortably in the home of his friend Chanut, the French ambassador, whose family afforded him all the comforts of French culture in a very rude and semi-civilised society. The Queen treated him with respect, and was evidently impressed with his great ability. She gave him leave to be excused from all Court ceremonies, and only to visit her privately when she required his instruction. His leisure was occupied in composing a masque or pastoral allegory in prose and verse, to be performed at the festivities which celebrated the Peace of Münster. Baillet, who had read the manuscript, calls it a *Fable bocagère,* in which the love of wisdom, the search after truth, and the study of philosophy, were set forth under the disguise of his dialogue. But we may be sure that his naturally dry and methodical discourse, which Queen Christina thought so superior to the paltry arguments of other men, did not produce any remarkable poetry, and his executors seem never to have thought of publishing it.

~~THE CROWN OF THORNS~~

THE CROWN OF SWEDEN

And the scholars platted a crown of laurel, and put it on his head, and they put on him an academic gown.

PRESENTLY THE QUEEN began serious study with him, and required him to be with her at five in the morning several days of the week, in the November of the Swedish climate. Nothing can have been more detestable and disturbing to a man who had all his life, and in better weather, lain in bed till eleven in the day, meditating and writing. But he was a courtier and obeyed. She next urged him to set in order his papers, and arrange them for publication; and here again he obeyed, most fortunately for his literary remains. But his care was almost rendered nugatory by the accident that upset the boat which was bringing all these papers up the Seine to Paris, where Clerselier had undertaken to edit them for Chanut. After lying in the river for a couple of days, the box was recovered, and the sheets of paper dried on lines like clothes by servants, thus disarranging their order, and producing loss and confusion. This happened in 1653.

The Queen was anxious to retain him by all means at her capital, and there is some evidence that she consulted him on politics, though his influence was certainly never sufficient to make her take the least notice of the Princess Elisabeth. But she desired to found a learned Academy, and asked him to draw up its rules, which may still be read in Baillet's *Life* (ii. 412). They are simply the laws of orderly discussion, each member in turn proposing a problem, and the rest discussing it in their order. The Queen was supposed to preside, and the 9th rule provides that "when it shall please her Majesty to finish the meeting, she will favour those present by giving a complete resolution of the question, praising the reasons of those who shall have approached nearest to the truth, and changing and adding what may be necessary to make it perfectly plain." This was indeed a royal task! He was also so impressed by the evils resulting from the jealousy of foreigners, that he expressly excludes all but natives from membership, giving the Queen only the right of asking foreigners to be present, and to offer their opinion if specially invited, after the members have spoken.

He left the sick-bed of his friend Chanut to bring these rules to the Queen on the 1st Feb. 1650. The ambassador had been attacked with inflammation of the lungs, owing to the terrible severity of a winter exceptional even in Sweden. Descartes was suffering visibly from this rigour of temperature, and the Queen, who saw that the climate did not suit him and would drive him back to Egmond, was planning to endow him with a large estate in the Duchy of Bremen, or in Pomerania, which she had acquired by the recent peace, so that he might remain her subject, and yet enjoy a country more suited to his constitution. But, with that thoughtlessness and selfishness so common in royal personages when the comforts of others are concerned, she forgot in her ambitious prospects that she was daily threatening to defeat them by her audiences at five o'clock in the mornings of an arctic winter. The rules he brought her were the last product of his pen, and his visit on the 1st February the last he ever paid his patroness.

~~THE DEAD CHRIST~~

THE DEATH OF DESCARTES

Then took they the body of Descartes, and interred it in a neglected graveyard, as is the manner of the Protestants to bury foreigners.

NEXT DAY he was seized with the very ailment from which his friend was but slowly recovering. He soon became delirious, and refused to submit to the treatment which had apparently succeeded with Chanut. The Queen's first physician, a Frenchman and a friend of Descartes, M. Reyer, was most unfortunately absent, and no other Court physician could be obtained than a certain Weulles, a Dutchman, who had sided against Descartes in the Leyden controversies, and, being a stanch Peripatetic and adherent of the old theories, was a declared enemy of Descartes at the Swedish Court. The very sight of this man set the patient beside himself. When they attempted to bleed him—an operation which he always discouraged, and on which he had far more advanced views than his opponents—he cried out in agony, *"Messieurs, épargnez le sang français!"* In calmer moments he told Weulles he would die the more peaceably if he would keep out of his sight. In an extant letter, this physician gives an account of these things, and says that, from his very first visit, he saw the obstinacy of the patient, and the violence of his fever, rendered the case hopeless. On the seventh day he became calmer, and being persuaded by Chanut, who was now well enough to help in nursing him, he allowed himself to be bled twice, but without diminishing the fever. From that time onward he discoursed reasonably and with great calmness on his impending death, and was so easy on the ninth day (Feb. 11, 1650) that he asked his faithful valet, Henry Schluter, to take him up and dress him. But presently he fainted, was again laid in bed, and died in a few hours, surrounded by the affectionate Chanut family, and fortified duly with all the rites of the Catholic Church.

Thus died the great philosopher of his age, without having completed his fifty-fourth year. The sorrow of his friends and the disappointment of his admirers may well be imagined. His remains were at first buried in a Catholic cemetery at Stockholm, then removed in 1666 to Paris, when they were interred with great pomp at

Saint-Geneviève-du-Mont, though the *éloge* prepared for the occasion was at the last moment suppressed by the Court. A mask of his face was taken after death for the Queen of Sweden, and pompous epitaphs on the four sides of his tomb recorded his virtues and his renown.

~~THE BURIAL OF CHRIST~~

THE REBURIAL OF DESCARTES IN PARIS

Now in the place where he was reburied, there was a Pantheon; and in the Pantheon a new necropolis, wherein was never man yet laid.

WITH THESE EXTERNAL marks of his greatness we are now little concerned. It is of more importance to note how the expectations of his followers—that he would presently have astonished the world by new discoveries—are belied by the study of his works, and by what we know of the many unfinished manuscripts, most of which were printed after his death. It does not appear that for many years—in fact, from the first publication of his *Essays*—he had ever abandoned one jot of his theory, or reformed even the most insignificant of his views. He was perfectly convinced that his *a priori* construction of the world was mathematically demonstrated, and had he lived years longer, he would certainly have done nothing more than elaborate it after the manner of his theory on the passions, and of the formation of the foetus and the structure of man. The experiments which he was always projecting were intended to discover nothing new, but merely to confirm his published theories. In mathematics he might no doubt have solved new problems, but this study he had abandoned in order to devote himself to physics, medicine, and ethics. He was determined to accommodate every phenomenon to the principle that all matter was extension, and all physical qualities a combination of figure and motion. As he never had once seriously confessed himself in the wrong, it may be regarded as morally impossible that with increasing age he should have entered upon new paths.

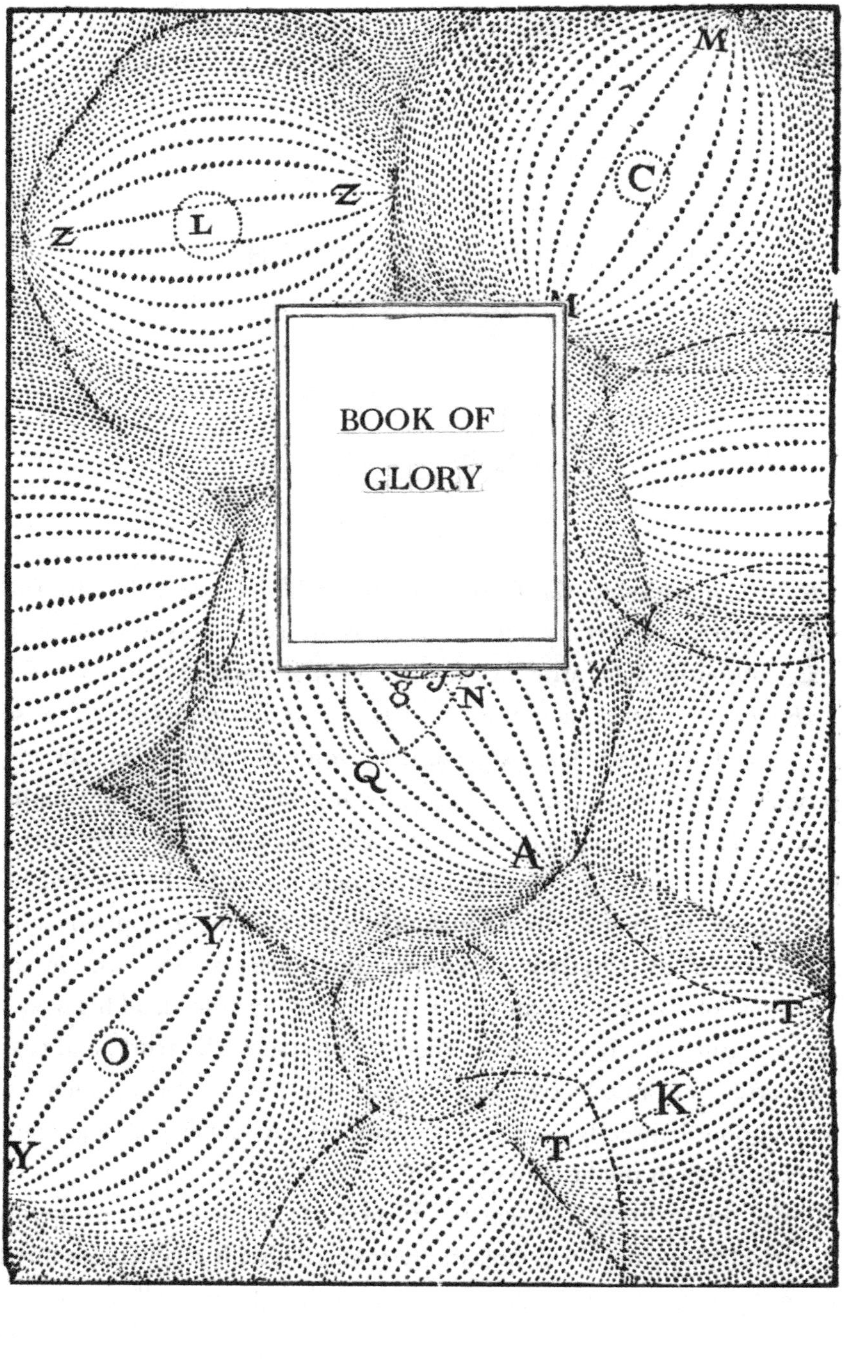
BOOK OF
GLORY
M
C
Z
L
Z
N
Q
A
Y
O
Y
T
K
T

9
DESCARTES' PHILOSOPHY AND METHOD

It is remarkable that Descartes has nowhere expressed himself so clearly concerning his Method as in the tracts drawn from obscurity long after his death, and first published in 1701, the *Recherche de la vérité,* and the *Régles pour la direction de l'esprit.* These *Régles*, as has already been shown, date from his early career, and are probably the first rough sketches thrown aside or worked up in the *Essays*. They are particularly important because they have been persistently ignored by the English expounders of Descartes, and by M. Cousin, in their desire to make the father of modern philosophy the founder of a system based on consciousness—that is, on an empirical observation of the mind. These tracts—indeed, all his works—show quite clearly that this is the very reverse of the truth, and that Descartes' was a deductive system, based on mathematical principles, and opposed directly to the Baconian, which consisted in a mere systematising of the facts of experience.

We know that Descartes turned aside with a profound dissatisfaction from the Scholastic system. His Jesuit teachers assumed from Aristotle, from St. Augustine, from St. Thomas, all manner of principles which they did not clearly understand, and did not verify for themselves. They regarded syllogism as the great engine for expanding and enlarging these principles into a system of knowledge. Descartes proposes to reject all such principles derived from mere authority; and he perceives that syllogism, however useful for expounding and explaining, is idle for the purpose of adding to knowledge. What we want is, first, clear principles, and then a process of deducing from them sure consequences. We seem to have a clear specimen of the latter process, at all events, in mathematical Science, which is occupied, indeed, with the mere surface of things, perhaps with mere imaginary forms, but which is able to deduce with certainty from its first principles, and build up a just system of knowledge, contrasting broadly with the Scholastic philosophy, first in its clearness, and secondly in its progressiveness. All other Sciences compared with this are mere history or polymathy, and produce intellectual indigestion, not growth in knowledge. But not only is the method of mathematics true and fruitful, it is the only true method; for all Sciences combined are nothing but the human intelligence, which is always one and the same, however various its objects. It is surprising, he thinks, how many men apply themselves to discover the properties of plants, the movements of the stars, the transformation of metals, and suchlike things, while hardly any one occupies himself with the human intelligence, or with that universal Science just mentioned. Truth, then, in its strict sense, must depend altogether on the knowledge of this Science, indicated by mathematics. We almost imagine we hear Kant speaking. Let us reject all probability, and therefore all reasoning from experience, which is often deceptive, and declare as a principle that those who seek the straight road to truth must not occupy themselves with any object of which they cannot be as certain as they are in the demonstrations of arithmetic and of geometry. What is the form of these demonstrations? Perfectly clear and simple ideas as starting-points, and these perceived by intuitions, by the simple inspection of the mind. Perfectly gradual and evident deductions from these intuitions, as we see them in Euclid. If, then, a complicated problem is given us to solve, we must first parcel it out into simple ones, by a careful enumeration, which Descartes calls induction, and taking each simple intuition, reason from it by strict deduction. What is required, next, to enable us to apply this infallible method not merely to a few problems about simple figures, the outlines of visible things, but to all the departments of human knowledge? First, to make our mathematical reasoning general, to get rid of the difficulty of complicated figures, which weary and perplex the imagination, and to make our proofs merely logical by the use of symbols. This was done by the discovery that geometrical truths could be expressed by algebraical symbols. Second, to find out the simple intuitions in nature which will serve as the safe starting-point for similar deductions in more fruitful Sciences. If, above all, we could find out the laws of applying pure mathematics, the Science of lines and numbers, to the qualities of bodies, which seem so totally distinct from them, then our method would be complete. And yet as Science is one, and its method is uniform, this application should be possible. By the principle that all matter is only extension, and that, therefore, all its laws are merely laws of figure and motion, this application is secured, at least in physics. Such is the plan of the early and unfinished *Régles*.

IN THE *DISCOURSE ON METHOD,* on the contrary, the philosophy of Descartes starts, as everybody knows, from universal doubt in speculation. He reflects the temper of an age weary of the old and the effete, despairing of knowledge from the received methods, and already awakened to independent thought by the reformation in religion. The fall of the Scholastic philosophy was certain. Not only with Bruno, Vanini, Campanella on the Continent, and with Bacon in England, but even in Dublin, we find works appearing to show that the current philosophy is confused and false, and must be replaced by a new system—*jam conducendo loquitur de rhetore Thule.* The contempt of antiquity became a ruling fashion among thoughtful men; and, as they generally loved to express it, they must no longer be content with the crude attempts of the real youth of the world, when they had now attained to older centuries of maturer discretion. We, proclaimed Bacon and Descartes in common—we are the true ancients. The Greeks and Romans were in Science but children that walked with slow and uncertain step. Let us then cast away their stale methods, now worn threadbare, and patched by the abuse of countless pedants, and begin afresh. Let us purify our intellect from all prejudices; let us get rid of all books, and see what the light of reason will teach those who use it with unshaded lustre.

But lest men should suspect Descartes of scepticism in the theological sense, which was then held the most odious of crimes, and an outrage to society, he carefully guards himself by excepting religion and morals—the one, an affair of inspiration, taught by Divine authority; the other, the cement of society and the bond of social order—from the invasion of his doubt. The former he declines altogether to discuss, submitting himself to the Church in which he was born. The latter he accepts provisionally, in the general form agreed upon by civilised society, and by the experience of civilised men, as a temporary residence, which is absolutely necessary when a man undertakes to pull down his house and build another upon new foundations. This reservation as to morals was well enough, but the theologians saw quickly enough that the severance of philosophic from religious scepticism would not last long. The absolute right of private judgment in philosophy could not but assert itself in other and very kindred branches of inquiry. Descartes might be a Catholic, a Conservative in faith, a pet pupil of the Jesuits; he was nevertheless in temper a Protestant, a sceptic, the spiritual father of Spinoza.

Having therefore established as his starting-point universal doubt—of the external world, of God, of our bodies, of our very minds as things known and understood—is there any possibility of finding any new and clear foundation from which to build up our knowledge? Everything is gone—mind, matter, Science, experience; all is, or may be, delusion; nothing remains but doubt. How then can we find a fresh starting-point? Evidently in the fact of doubt alone. What is doubt? A state or condition—in fact, a judgment; and how can there be a judgment without someone to judge? Doubt, then, is an act of thinking. Thinking is inconceivable without a person to think. Thus, doubt implies the mental existence of a doubter. If we had not started from this act of thought, any other would serve us as well. Walking, eating, remembering, any conscious act, will afford us the same conclusion. *Cogito, ergo sum.* It is no syllogism, no formal conclusion from a single premise. It is an inevitable and immediate inference.

In spite, however, of Descartes' attempt to abandon all established principles, they reassert themselves at every step of his system. He himself, in his later and more Scholastic *Principles*, bases this inference from thought to existence on the principle of Contradiction. In any case it must presuppose the brocard often paraded by him, that "nothing can have no qualities," or that a quality is a manifestation of something beyond itself—at least, of something more general than itself. This is the principle of Causality.

THE *I* WHICH THINKS, and whose existence is thus established, has nothing to do with matter, or the external world, or even the body, none of which belong to our idea of it. It is simply a thinking substance, whose very essence, so far as we know, consists in thinking, and in thinking alone. The conclusions thus attained are not only valuable in themselves, but lead us to further inferences. How and why have we attained this certainty? Simply because we perceive it so clearly and distinctly that its denial is impossible. Hence follows the first and fundamental rule of Descartes' philosophy, *to accept as true what we perceive clearly and distinctly,* and nothing else. But from the very outset he saw that there was difficulty in discovering and determining what we really so perceived, and what, on the other hand, was thrust upon us by tradition or prejudice. He himself, though he often approaches the capital distinction between immediate intuition and thought, or mediate inference, never lays it down as a principle.

He believes inferences, not only mathematical, but metaphysical, if clear and distinct to his mind, are as certain as the *cogito, ergo sum.*

We can hardly doubt that in its original form his system ought to have established extension on the same basis as thought, being the clear and distinct perception which we have of a quality different from thought. But Descartes' philosophy was the very opposite of what historians of philosophy have described it—it was not a system based on the observation of the facts of consciousness. It was a deductive system, drawn, as a mathematician would be sure to frame it, from the fewest possible assumptions. Hence he reaches the external world, not by clear and distinct perception, but by a roundabout inference. For he determines in the first place to deduce from the fact that we exist as thinking beings the existence of God.

This he does in his various works, according to three different demonstrations, which he himself declares to be in reality identical, but which again he says *(Reply to first objections)* may be classed under two kinds—the argument from effect to cause, and the *a priori* argument of explicating the existence of God from the idea He has given us concerning Him. Let us follow out this latter hint, and state the arguments in order. First comes that in the *Discourse on Method,* which immediately follows the proof that we exist as thinking beings. If we exist as such, we have thoughts, and these thoughts may be regarded either as mere mental objects, or as the consequences of causes independent of us. In the latter case, the cause contains formally what is contained in the idea objectively. Now comes in a curious metaphysical assumption, which again shows how hopeless was Descartes' attempt to get rid of all the principles of his teachers. He holds it to be obvious that no effect can be more perfect than its cause, and that consequently the cause of any idea must contain, formally, at least as much as the idea contains objectively. But I have within me, and everyone has (another assumption), the idea of a perfect—viz., of an all-wise, infinite, omnipotent Being, contrasted with the idea of myself, in whom the very fact of doubting proves imperfection. This perfect idea, then, we cannot have produced, as we may have produced other ideas, nor can it be derived from any finite object known to us. It must accordingly have been produced by a real Being containing formally all these objective perfections. This, then, is the Deity. We know Him not intuitively, but by necessary inference.

But if this proof should be considered too abstract, he gives it in another form, which he thinks more comprehensible, and which derives the existence of the Deity from the very fact of our own existence. If we are not created by God, only three other hypotheses are possible—either we are created by ourselves, or by our parents, or by some other cause less perfect than God. The first is impossible, or we should not have created ourselves with the imperfections, the wants, the unsatisfied aspirations which we possess. The second only moves the difficulty one step backward, and in any case we have in us the idea of all perfection, which could only be produced by an adequate cause. Here we come back to the former proof, which disposes of the third alternative. Thus the very fact of our existence—nay, of our existence at any moment—presupposes a perfect Being who has created us, and keeps us in existence during each moment by his interference. These are the two forms of the simpler kind of proof, which argues back from acknowledged facts to their only adequate causes.

The third, or synthetical proof, which he calls geometrical, starts from the fact that we have in our minds various ideas, not perhaps given by sense, or existing without us, and nevertheless containing fixed and immutable properties, so that we cannot acquire the idea without necessarily holding the truth of these properties. Thus I cannot form the idea of a mountain without a valley, and I cannot understand the nature of a triangle without admitting that its angles are equal to two right angles. Now, on analysing the idea we have of God, as the most real Being, containing every perfection, I find that existence must be comprised among these perfections; otherwise the idea could be enlarged by adding this quality, which is absurd. In other ideas existence is not such a necessary ingredient, because they are not considered as unique and absolutely perfect. But in this, existence must be comprised. Hence the very idea of God, rightly understood, includes in it necessary existence, and the existence of the Deity is proved from the very fact that we possess a notion of Him. This is the argument at the basis of all his demonstrations.

EVERY ONE KNOWS what a tempest of controversy this demonstration evoked. It was attacked as a mere copy of the Scholastic argument of St. Anselm; it was attacked as assuming the universality of the idea of God, whereas many nations of savages did not possess it; above all, it was attacked as tending to give absolute reality to our subjective convictions. Nevertheless, the proof maintained itself, with the help of Newton, Clarke, and Locke in England, and the Cartesians in France, in spite of much opposition, till Kant's day, when the most conclusive part of the great *Critique* was directed to prove that existence was no quality attaching to an idea, which added to or detracted from its perfection, but a peculiar relation to our thinking or to our experience. Hence no manipulation of a mere idea could ever extract existence from it.

Nevertheless, the principle of drawing a proof of existence from thought seemed to be sanctioned by Descartes' very first step, *cogito, ergo sum;* and so Hegel, whose whole system is based on this principle, has so far rehabilitated the demonstration, that we see it now reappearing among Hegelian theologians, and Principal Caird's last book shows us a curious revival of it.

From the demonstration that there exists a supreme Being, endowed with all perfection, among which are included unity, benevolence, veracity, omnipotence, there follows at once a fresh guarantee of the certainty of all our knowledge. It is indeed true that we had already found that clear and distinct perception was the primary test; but what if a malign creator, a spirit of evil, had chosen to deceive us, and give us clear perceptions of idle dreams? It is only on the supposition of the veracity of the Deity that we can be certain of the knowledge received through our faculties. Hence, says Descartes, an atheist can have no true Science. He is said to have had Montaigne and Charron in view here, but it is far more likely that he was thinking of the Greek Pyrrhonists, who doubted even the truths of mathematics. For when his adversaries accused him of reasoning in a circle—of establishing the existence of the Deity by our clear and distinct perceptions, and the truth of these perceptions by the character of the Deity—he always answered by saying that a clear and distinct perception is always at the moment the only infallible test of its own truth. But as we cannot go back upon our previous acquisitions, and hold in view the steps by which we attained them, we must here depend on some general principle of the stability of knowledge, the immutability of the same relations between immutable things, as Locke puts it, and this is derived from our conviction that there is a fixed order in the universe, which is not capriciously changed. In fact, if the Deity be regarded as the order of the universe, it will be merely a theological expression for the Permanence of Natural Laws. But as Descartes put it, he did not avoid the reasoning in a circle. It is surprising that he did not sever subjective and objective certainty, in the modern sense, and base the former—our conviction—on its own clearness; the latter—its guarantee—on the existence of a Divine order of nature. But granting his argument, the knowledge that we are created by such a Deity encourages us to proceed to other clear perceptions than that of our thinking selves, and to draw from them, by strict deduction, conclusions which will have the guarantee of being universal and certain knowledge.

10

HIS SYSTEM OF PHYSICS

Passing on, then, from our minds, already given us as thinking things, and—since they are the subjects of the attribute of thinking—as thinking substances, we find our minds closely connected with bodies; so much so, that many of our thoughts are not conceivable except as produced through our bodies. Here, then, is a kind of substance, not thinking, and so far heterogeneous from mind. What is its distinctive attribute? We can imagine it without hardness—for suppose it were created so as to remove whenever we approached it, we should never have known it as solid; or without weight—for we see some bodies, like fire, which have very distinct qualities, but apparently no weight. The same goes for properties of colour, odour, etc. But one thing we cannot abstract from bodies—their extension. This extension, which has, and can only have, three dimensions, is so eternal and certain an attribute of matter, and so clearly and distinctly known, that the clearest and most distinct Sciences—mathematics—are constructed upon it. The laws of extension, in the form of figure, are the Science of geometry. The laws of the measurement of extension, and of the motion from one part of extension (place) to another, are the Sciences of arithmetic and mechanics. Extension, then, and not impenetrability, as is often supposed, is the essential attribute—to us, the essence of matter. And as the Science of geometry determines the laws of its figures, and the use of symbols enables us to deduce conclusions with certainty, even when the figures are too complicated to be represented, it follows that, by the study of mathematics, we can unlock the secrets of nature, and learn the composition, the properties, and the mutual relations of bodies.

This, then, is Descartes' great physical theory, and one which was far more important than his mathematical discoveries in the history of Science. It is the theory that, instead of assuming specific differences of quality in matter—instead of assuming substantial and accidental forms, and their mutual cohesion and expulsion—instead of assuming an imaginary void to account for rarity and density—starts from the intelligible hypothesis, confirmed in strange ways by modern experiment, that matter is homogeneous, that it is conterminous with extension, and that all differences of quality are simply produced by a different mechanical composition, and a difference of motion in its parts.

"Give me extension and motion," Descartes exclaims in his *Monde*, "and I will construct the world." The primary qualities of matter were plainly enough modifications of these. It required Scientific boldness and insight to assert that colour, sound, heat, and all the other secondary qualities were produced in our sensibility in no other way. Thus he anticipated the mechanical theory of the transmission of light and heat; and had he known that an appreciable time elapses during the process, he would have been strongly confirmed in his *a priori* conjecture. The production of sound he perfectly understood and explained. His denial of a void, and his assertion that beyond the atmosphere there must be some subtle substance, because there is extension—this theory was not verified till quite recently, and by the delay of the movement of bodies through what was supposed to be empty space.

THUS, then, Descartes set about the construction of a universe *a priori,* through the action of motion impressed on infinitely extended matter by the Creator. The infinity of this matter, or more strictly its non-limitation, either in space or time, he firmly adopted. Of course, he added, we know by revelation that the world was made in six days, and by direct creation of the Deity; but it peculiarly satisfies the mind to comprehend how these things could take place by natural and gradual means. He assumes that homogeneous matter became separated according to the minuteness of its component parts, and the amount of motion consequently varying, into three large divisions: first, the most subtle and volatile kind, which makes the heavens and the upper ether; then the next in degree, which makes our atmosphere, and fire; lastly, the most solid kind, which became earth, metal, and other hard substances. The parts of the subtlest kind are far minuter than the minutest particles of dust, and capable of constant and violent motion. When this was once established among them, the displacement of particles could only happen, in a plenum, by others taking the place they had left, and thus a circular motion, such as we see in eddying water, is the universal consequence of the direct or natural motion of the universe. This reminds us of Plato's *Timæus*.

Thus are produced an innumerable series of vortices of matter, of various volume and various regularity of form, in which are carried along the grosser bodies situated in them. Our solar system is such a vortex, and the earth, though in one sense moving along with the matter of its vortex, in another is at rest, as the passenger is at rest in a moving ship. The easier and quicker motion of the subtler matter in each vortex causes the grosser to deflect towards the centre, and this is the principle by which he explains weight and gravity—a hypothesis regarded with admiration by his followers. Solidity he explains as nothing but an absence of motion in the particles of a body, and he can conceive no stronger "cement" which could hold them together.

We need not follow into greater detail this theory of general physics. It was elaborated by Descartes as regards minerals, and he gave a special explanation of the growth of the human body. But the explanation of plants and of the lower animals was still wanting when he died. The vortices of Descartes have been since Voltaire's day decried on account of the greater simplicity of Newton's explanation of the solar system. But when we extend our view to the universe, and to the genesis of systems like our own, we cannot but respect this great hypothesis, which reminds us of the theory of Laplace. But in its detail, and in the more special laws of motion which its author supposed, it was false, and was soon abandoned as obsolete. Still it is to Descartes that we owe the first attempt to explain mechanically our solar system, the formation of planets, the relation of the tides to the moon, and the attempt to subject the laws of motion to Scientific analysis.

THE EARLIEST ESSAYS on this mechanical theory of the universe were the tracts on optics *(Dioptric)* and on *Meteors*, which he published in his first volume. The perception of light appeared of all sensations that least to be explained by motion and by ordinary contact of particles. His law of ordinary refraction, and his application of this mathematical principle to explain the rainbow, were a momentous onward step in the rational explanation of physical phenomena, as opposed to the Scholastic plan of assuming occult qualities and occult action. The general principles of both reflection and refraction of light he understood perfectly well; and these were not vitiated by his two mistakes—first, that the action of light was not by vibrations of the ether, but by the propagation of direct motion from one minute particle of subtle matter to the next, and so on in right lines, till the last of the series affected the eye. This was at least better and more philosophical than to hold that the same particles which started from the distant object reached the eye. Second, he was not aware that any time elapsed during the propagation of light, and compared it to the action by which a blind man feels with the end of a stick. But it required long experience and investigation, based on the very improvements in optical glasses which Descartes described and directed in his *Dioptric*, to expel these erroneous views from the minds of the most Scientific students. How far he was in advance of his day may be seen from the 6th Discourse, in which he explains the perception of distance, and lays down explicitly all the arguments and illustrations used long afterwards by Berkeley in his Theory of Vision. It is impossible that Berkeley can have been ignorant of Descartes' *Dioptric*, and yet how he could claim any originality whatever on the subject is beyond strange. The convergence of the optical axes, and how this may be supplied by successive observations with a single eye, the varying colour of the objects, the greater dimness, the number and kind of intervening objects, the uncertainty of all these various indices —all this, which Berkeley urged, is found in Descartes' *Discourse*; nay, even the illustration of the moon looking larger near the horizon than when high in the heavens.

THE MOST INTERESTING Scientific anticipation in the *Meteors* is that in its 8th Discourse, when he treats of the nature of colours. He was the first to declare that difference of colours arose not from any specific quality in the objects, but merely from the varying velocity of the motions which affected the eye.

> In all this reason agrees so well with experience, that I do not think it possible, when we know both, to doubt that the thing is not just as I have explained it; for if it be true that the sensation we have of light is caused by the motion, or inclination to motion, of some matter which touches our eyes, as many other things show, it is certain that diverse motions in this matter must produce in us diverse sensations, and as there can be no other

> diversity in these motions than that explained, so we do not find by experience any other in the sensations we have of them than that of colours. ... It is nonsense to talk of some ideas being true and some only apparent or false, as the philosophers say; for all their true nature being only to appear, it seems to me a contradiction to say that they are false and yet appear.

Accordingly, he assumes in the particles acting on our eyes not only direct but also rotatory motion—the latter increased or diminished by striking on surfaces at various angles. If the rotatory motion is very rapid in proportion to the direct, red is produced; if less rapid, yellow; if the rotation be slower than the direct motion, blue and green. These gradations he inferred from the order of colours in the rainbow. For in his *Meteors* he approached special phenomena, like the rainbow, and rare phenomena, such as those of *parhelia*, and showed how all these were the result, not of special interferences with the laws of nature, or of any new law, but rather of rare combinations of the ordinary causes of phenomena, which thus produced exceptional results. Here, again, we have the rational and the orderly explanation of Science opposed to the guesses of metaphysicians and the superstitions of the vulgar. The principle of Parsimony is everywhere openly proclaimed by Descartes. Since we know that the minute composition and movement of particles of matter varies the sensations produced by them in us; since we find that the same sense of touch produces in us such opposite sensations as pleasure and pain, pressure and titillation; since both sensations of light and of sound can be produced in us by violent impact; and since we know of no other kind of action in matter which produces sensations—we are warranted in assuming that all the action of bodies upon us is merely some kind of contact, differing according to the figure and the motion of the particles which touch our organs.

But when, in pursuit of this principle, he gives *a priori* explanations of all manner of phenomena in heaven and earth, deduced from the motion of diverse particles, he confesses that the plurality of causes which may produce the same effect is his great difficulty. He feels that endless time and outlay is required to verify his theories by crucial experiments; and till that has been done, he can offer nothing but the satisfactoriness and simplicity of the explanation as a guarantee of its truth. Once, indeed, he advances the statement, that the veracity of the Deity would come into question if He permitted us to be deceived in following such strict and sober demonstrations. [1] But in general the distinction of purely mathematical and of physical proof is acknowledged by Descartes, and he confesses the possibility, though he never admits the fact, that he might be mistaken. But even if this were so, he holds that to provide a rational solution is the main thing, for it indicates a general possibility, and therefore proves that, even if mistaken, it would be replaced by some solution similar in kind, and derived from the same principles. Hence the vital importance of showing that the phenomena of nature can be explained mathematically. This hypothesis is so much simpler, more complete and Scientific than any other, that if it be proved possible, it cannot but be accepted as true. Thus it is very unfair to Descartes to estimate the value of his work by the strictly accurate discoveries which he made. For surely Descartes' greatest claim to the gratitude of posterity is not to be found in his actual solutions of problems, but, as he himself insisted, in his anticipations of the *conditions of the right solution* in problems of physical Science. Even those who refuted his theories, and supplemented them by newer and simpler solutions, were led by the general light which he gave them, by the discovery he had made that nature could be interrogated mathematically, and the result set down in mathematical formulae. Thus it was no accident that Huyghens, educated in Descartes' principles, and under his personal influence, should have originated the now adopted vibration theory of the transmission of light. It is this new way of looking at nature which he forced upon the world, not less upon his opponents than upon his followers, and which is decidedly the greatest and the most fruitful of all his discoveries.

11

DESCARTES' PSYCHOLOGY—THE NATURE OF ERROR

The physics of Descartes were thus elaborated earlier than the other parts of his system, though the affair of Galileo delayed their publication. It is convenient to dispose of them in the same order, that we may turn at greater leisure to his psychology and ethics, and to the remainder of his metaphysics. His physiology is so closely connected with his ethics, that it is necessary to discuss them together, and not (as might be expected) in an appendix to his *Physics*.

We have assumed that what we know clearly and distinctly must be true, and that what is deduced from such data must be true also. This is based ultimately on his doctrine of *innate ideas,* which has caused much comment and many unfair refutations. For he explains in his letters that by innate ideas, he does not mean ready-made ideas, complete images, or pictures, in the mind of the infant. He means that the mind infused by the Deity into every human body has certain natural predispositions which compel it to adopt certain beliefs, as soon as it begins to reflect, and to exercise its faculties. Such are the ideas of God, of substance, of unity, and a host of others, which he never essayed to enumerate. Some of these might remain unsuggested for a long time, perhaps altogether, but whenever an occasion arose, they would appear and be assumed as indisputably true. It is in the case of these that the veracity of the Deity is of vital importance, for a malign Creator could have made us believe innate falsehoods.

But is there no difficulty, is there no controversy, as to what we do know clearly and distinctly? Is error not possible? Do not most men mistake early prejudices and false inferences for clear and certain truth? What, then, is the nature of error, and how is it to be distinguished accurately from truth? This is a necessary, and indeed a vital question in the Method of Descartes. He says in the 4th Meditation:

> When I come to examine the cause of the many errors which are manifestly made by human nature, I find that, together with the idea of a Being of sovereign perfection, I have as its opposite a negative idea of non-being *(le néant)*—that is, of what is infinitely removed from all perfection; and that I am, so to speak, intermediate between the sovereign Being and non-being, that there is nothing in me which can lead to error, insofar as the sovereign Being has produced me. But if I regard myself as participating to some extent in the *néant* or non-being —viz., insofar as I am not myself the sovereign Being, and that I am deficient in several things, I find myself exposed to an infinity of deficiencies. And thus I know that error, as such, is nothing real depending on God, but only a defect; and that to err I require no particular faculty given me by God for that purpose, but it merely happens that I am deceived because the power given me by God to discern truth from falsehood is not infinite.

This is one of the numerous passages which Descartes honestly believed he had drawn fresh from his own speculation, but where the echo of his studies of Plato, Augustine, and Campanella unconsciously influenced his mind.

IF WE INQUIRE MORE CLOSELY how individual errors are to be explained by this general theory, we find that an analysis of our thoughts into classes becomes necessary. Descartes lays this out in the 3rd Meditation:

> Among our thoughts, some are as it were the images of things, and to these only is properly applied the term *idea*, as when I have before me a man, a chimera, heaven, an angel, or even God. Other thoughts have a different form, as when I wish or fear, affirm or deny; then I conceive, indeed, some thing as the subject of my mental action, but I also add something else by this action to the idea in my mind; and of this kind of thoughts, some are called *volitions* or affections, and the rest *judgments*. The mere perception of ideas cannot possibly contain any error; it is in our judgments concerning them that error consists. Thus I infer from these ideas that they are produced by external objects like them, because I fancy I am so taught by nature, and because they do not depend upon my will. And yet these inferences may be false. For being *taught by nature* means not only the evidence of that *natural light* which is the highest and most perfect guarantee of the truth of our simple intuitions; it may also mean a certain *spontaneous inclination,* a *blind and rash impulse,* which certainly deceives me, for example, in the choice between virtue and vice, and therefore cannot be trusted in the distinction of truth and falsehood. Thus our ideas might be produced by no external cause, but by some as yet undiscovered faculty within ourselves; and even if they were, this external cause need not resemble our ideas. Nay, in many cases we know that it does not. It is only by

> reflecting carefully on the truth, revealed to us by natural light, that all ideas of mental objects must be derived from causes which contain formally all the reality possessed objectively by the ideas, that I am able to deduce this conclusion: All the ideas of body which are clear to my mind—viz., extension, figure, place, movement, substance, duration, and number—are real and true; those of light, colour, taste, heat, cold, etc., are so obscure and confused, that nature teaches me nothing about their reality or their causes. They may even proceed from non-being, or from some want in my nature. And so of many other ordinary prejudices which have infected not only common life, but even previous philosophy.

It remains to analyse more closely the exact nature of this blind or rash impulse, which Descartes distinguishes from the natural light of reason. We may repeat that he has given no real test for severing them, especially as he considers the natural light not an immediate intuition, but often only attained by long reflection. His adversaries might readily differ with him on its application, and what one calls natural light, the other might consider mere blind impulse.

In the first place, error cannot exist in the bare representation of an idea. Each mental fact as such is a mental reality, about which there is no dispute. It is in the inferences about our ideas that we err. Inference implies judgment, and it is only our judgments which can be called true or false. But judgment implies two faculties—first, the understanding, which grasps the two ideas, and their relation; second, the will, which chooses that they shall be affirmed of one another. The main difference between these faculties is, that while the Deity has given us a limited understanding, unable to comprehend a vast number of things, our will is perfectly free, and apparently confined by no bounds. We have no right to expect from the Deity an infinite, or even a very large, understanding. We have no right to expect a free or uncontrolled will. The fact that He has given us the second is a privilege, which should not make us thankless as regards the first. This free will or liberty is not the liberty of indifference—a low degree of the faculty, which implies ignorance of the better and the worse motives. It means that when the understanding comprehends the motives, we feel that we can act upon them without the constraint of any external force. In the Deity there is no indifference, and the highest conceivable liberty. It is, however, in us the disproportion of the faculty of willing—of liberty—and the faculty of knowing, which produces all our errors. For the will proceeds to embrace judgments before the understanding has properly conceived, or when it is unable to conceive, the ideas required for these judgments. In such cases both affirmation and negation may be regarded as errors, because they rest upon no foundation. Hence our errors arise partly from our fault, partly from our defect. Our understanding is finite, and therefore unable to attain to perfect truth. Our will is rash, and induces us to beliefs for which we have no proper foundation. And yet it is because of this unlimited will that we may consider ourselves "made in the image of God," whose faculties have no limitation.

LET US INQUIRE, further, into what department of the understanding, into what sort of ideas, the will makes these incursions, and leads us into error. The understanding can think an idea of itself, by pure intellection; and here there is little room for error, for such *thinking* is eminently clear and distinct, it is accompanied with a great mental effort, and is only undertaken rarely, and by serious thinkers. Thus we can think a triangle, and form a clear idea of it; so we can think a chiliogon, and form a clear idea of its properties, complicated though it be. But there is another way of attaining these ideas—through the *imagination*; and here it makes a great difference whether the figure be simple or complex. A triangle is easily imagined, a chiliogon only very confusedly, and by an image which would answer equally well for a figure of 900, or any other large number, of sides. This imagination is then a particular kind of mental act, not necessary to our idea of the mind, for we can conceive thought without imagination, though not imagination without thought. It would also seem probable that were the mind joined intimately to a body, this imagination might be caused by an application of the mind to some suitable organ of the body. But the apparently corporeal nature of imagination affords no proof of the existence of bodies.

This imagining is, however, not confined to geometrical figures, but extends to all manner of qualities—heat, sound, taste, etc.; but as we perceive them more clearly by means of *sense*, we may pass to this third kind of mental action. This sense has been educated from childhood by our natural belief in our own bodies, in which we feel certain pleasures and pains by the application of sensations generally not under our control, and therefore assumed to come from external bodies. The importance, and hence the reality, of these sensations, is at first altogether determined by their being associated with pleasure and pain. Thus a hard body can hurt us by its contact, and hence is regarded as a more real substance than the air, which, because it yields to us always, even comes to be regarded as an empty void; and so of other sensations. Thus our minds are easily filled with prejudices crystallised in language, sanctioned and preserved by memory, which the least reflection shows to have no solid basis. We require the consideration that we are thinking beings, that there is a perfect God who has made us, and that what we perceive clearly and distinctly in bodies must come from some adequate cause, apart from His direct action, in order to understand how there is an external world, in accordance with His veracity. For He has given us a natural belief in an external world apart from Himself, and there must be some corresponding facts to this natural belief when it is explicated by clear and distinct thinking.

The most difficult instance against the Divine veracity which we can find to answer on this theory, is that of men desiring things decidedly hurtful to them, as for example the thirst of a dropsical patient. This is explained by the analogy of a watch, which does not the less follow just and general mechanical laws when it goes astray than when it goes right. We must take care to understand rightly the expression, "Nature teaches me this." The sense in which Descartes takes *nature* here is strictly human nature, composed of mind and body, and thus excluding both purely intellectual truths and purely physical qualities. Nature in this sense merely teaches us what to pursue and what to avoid for our comfort, and so far only is its teaching clear: beyond this we have no right to follow its suggestions without careful reflection. Nay, even in its own sphere, it only gives us the general law, and may be at fault in any particular case, as in that above stated. Descartes explains that each nerve produces only one kind of sensation; that a pain in the foot could be produced even after its amputation by action upon the nerves originally connecting it with the brain at any point up to the brain itself. Hence the sensation of thirst, naturally produced by a condition of the stomach and throat, and instructing us when to drink, may be produced by disease in the same nerves nearer the brain, and thus the patient is in one sense deceived. But he is deceived in consequence of the universality of a sound and useful law, which does its work in spite of a derangement elsewhere in the system.

12

ANTHROPOLOGY—THE AUTOMATISM OF BRUTES

We have now sufficiently discussed the nature of matter (in the physics) and the nature of mind (in the psychology) to approach the most difficult and interesting of all problems, that of the nature of man, as composed in a peculiar way of matter and mind, of soul and body, so that these are combined, though heterogeneous, into a natural unit. And first let us see how much of the phenomenon can be explained merely from the structure of the human body. On this question Descartes was making investigations for years. He was dissecting not only animals, but, as he once hints, human subjects. He even practised vivisection; and we have, through his various treatises and letters, accounts of such operations on rabbits, dogs, fish, and eels, for the purpose of watching the motion of the heart, the movement of the blood in the arteries, and other cognate questions. The theory which seems to have been popular in his day was that the soul, while consciously performing voluntary motion, unconsciously controlled the organs of the body, and was the proper cause of the beating of the heart, the digestion of food, the peristaltic action, and other vital functions. This is the theory which Descartes sets himself to disprove, seeing that mind consists in thinking, and that therefore it cannot perform actions without recognising them as such. The theory of latent modifications, which Leibniz established, was beyond Descartes' vision, and precluded by his theory of mind.

On the contrary, in his tract *On the Formation of the Foetus,* and *On Man,* as well as through several of his finished works, he develops a mechanical theory of the formation of the human body, like that of the other animals, starting from a fermentation produced in generation, which causes heat and expansion, so forming the heart, and next producing a motion of the subtler matter there found towards the locus which becomes the base of the brain, with a consequent return of the grosser matter into the places thus vacated. When this dilatation and contraction of the heart is once established by its heat, which can be felt by any experiment in vivisection, and when its dilatation is supplied by liquors flowing into it from the neighbouring parts, the human organism may be regarded as established. Vital functions are accordingly the result of heat and motion, just as mechanically as the going of a clock is the result of cog-wheels and pulleys. The phenomena of involuntary motion, such as starting, winking the eyes, etc., show that many bodily actions are produced in the same way, merely by the effect on our nerves of an external impression.

Here is his summary at the end of his tract *On Man.* Having shown that he has only assumed what has actually been seen by anatomists, and the existence of *invisibly minute structures* of the same kind, to account for what is yet unexplained, he concludes:

> I desire you next to consider that all the functions which I have attributed to this (animal) machine, such as digestion, the heating of the heart and arteries, nutrition and growth, breathing, waking and sleep, the perception of colours, sounds, tastes, heat, and other such qualities by the external senses, the impression of their ideas in the organ of the *sensus communis* and

> of imagination, the retention or impression of these ideas in memory, the internal motions of appetites and passions, and finally, the external movements of all the limbs, which follow so suitably as well from the action of objects presented to sense as from the passions and impressions which are found in the memory, that they imitate as perfectly as is possible those of a real man—I desire you to notice that these functions follow quite naturally in the machine from the arrangement of its organs, exactly as those of a clock, or other automaton, from that of its weights and wheels; so that we must not conceive or explain them by any other vegetative or sensitive soul, or principle of motion and life, than its blood and its spirits agitated by the heat of the fire which burns continually in its heart, and which is of no other kind than all the fires which are contained in inanimate bodies.

Thus the mere animal called man, and the other animals, are only skilful automatons, constructed by the Deity according to the general laws which He has impressed upon matter, and they admit of none but a mechanical explanation. Hence the lower animals, in which we have no reason to assume the infusion of a rational soul, as they use no language, and perform no actions which cannot be proved the direct result of their material organism, are mere animated machines, in whom the signs of joy and grief, of anger and fear, are merely signs of a motion in their animal spirits similar to that which is sometimes induced in us by external objects, without the participation, and against the judgment, of the mind.

WE HAVE YET to explain what these *animal spirits* are, which play so large a part in Descartes' physiology, and which held their place long after till the rise of microscopic anatomy. They were adopted from

Galen, whose *natural spirits* in the liver, and *vital spirits* in the heart, Descartes rejected. They still survive in our ordinary language. In introducing his discussion of the sense of sight, he there indeed tells us that the mind sees, and not the body, but this through the mediation of nerves, which concentrate in the brain, so that a brain disease may stop the intelligence which comes from any point of the extremities. These nerves, he says, are composed of three parts: first, the enveloping skins, which are mere prolongations of that which surrounds the brain, and may be compared to branching tubes (like the veins) throughout the body. Then there is the enclosed substance in the form of several independent filaments stretching from the brain to the extremities. Lastly, the animal spirits, which are like a very subtle air, and which, coming from the chambers of the brain, flow by the same tubes or canals into the muscles. These three substances, he says, are already admitted by physicians, but their respective uses have never been explained, especially as regards the two diverse functions apparently performed by the nerves—the one of announcing sensations, the other of producing motions. There cannot be separate nerves for these, as is often assumed; for no motion is ever produced without sensation, though in paralysis it is possible to have the former destroyed and the latter preserved. Descartes thinks that the nerves or filaments produce sensations, but that the animal spirits flowing into the muscles and expanding them produce motion. These animal spirits, moreover, keep the tubes of the nerves open and in proper order, and protect the enclosed filaments. On this theory the notion of images propagated to the brain is absurd: there are only nervous excitements, and these suggest to the mind the presence of objects corresponding in some way to them.

If we suppose a machine of this kind, with a thinking soul *infused* into it by the Deity at the first moment of its existence, then the soul, radically distinct from the body, and of an opposed nature, must be brought into contact with the body somewhere, that it may both receive impressions and direct motions. This must evidently be in the brain, whither all the nerves of sense converge; and in the brain, as most parts are double, we must place it in that portion which is, so to speak, the single centre of the mass, the *conarium*, or pineal gland. At this point, in some inexplicable way, and by the special arrangement of the Deity, the mind is specially in contact with the nervous organism. This being so, while a great number of actions and of passions are automatic, and the mere result of the organism, like nutrition, or the depression of hypochondria, others are produced by the special action of the mind on the pineal gland, and so through the vital spirits on the body. When the body decays and dies, there is no reason to think that the mind, which is immaterial, dies with it. Thus the immateriality of the mind leaves the way open to the theological arguments for its immortality, a question Descartes hardly touched, doubtless because the Church insisted on the immortality of the body also, which he must have regarded as absolutely a miracle, and naturally impossible.

Two points in this exposition seem to require a word more in explanation—the automatism of the lower animals, and the production of such passions as love and hate by merely mechanical causes operating in our organism. What is it that feels the passion? Is it the body, or is it the mind? If the brutes have no minds, how can they feel these passions? In the case of man, this problem leads us up to Descartes' ethics—the final point of his system, as he himself conceived it.

He was pressed on all sides by objections as to the intelligence of animals; he was supplied with endless anecdotes about their sagacity, and with comparisons of brutes with savages, in which the latter seemed the inferior. Between the highest man and the lowest, it was urged, the difference is far greater than between the lowest and the brute. Besides, the feeling of humanity is outraged by this "barbarous and cruel" theory, as More rightly called it; for we know that it led Descartes and his followers to a reckless indulgence in vivisection, merely to witness the internal structure of the animated machine, as we should take to pieces a watch. This was particularly the fashion at Port Royal.

THE ARGUMENTS by which Descartes met these objections were very able and logical—and in few of his theories does he show less dogmatism, although he never cedes his point. The two most important and explicit passages in his works are his letters to the Marquis of Newcastle and his reply to the objections of More. In these he argues as follows:

> As to the understanding conceded by Montaigne and others to brutes, I differ, not for the reason usually alleged that man possesses an absolute dominion over the brutes, which may not always be true, either as regards strength or cunning; but I consider that they imitate or surpass us only in those actions which are not directed by thought—such as walking, eating, and putting our hands out when we are falling. And people who walk in their sleep are said to have swum across rivers, in which they would have been drowned had they awaked. As regards the movements of the passions, although they are accompanied in us by thought, because we possess that faculty, it is yet plain that they do not depend upon it, because they occur often in spite of it, so that even their more violent occurrence in the brutes cannot prove to us that they have thoughts. In fine, there is no single external action which can convince those who examine it that our body is not merely a machine which moves of itself, but has in it a thinking mind, except the use of words, or other signs (such as those of mutes) made in relation to whatever presents itself, without any regard to the passions. This excludes the talking of parrots, and includes that of the insane, as the latter may be apropos, though it be absurd, while the former is not. It also excludes the cries of joy or pain, as well as all that can be taught to animals by acting on their hopes or fears of bodily pleasure or pain; which is the principle of all training of animals.

It is remarkable that language, so defined, applies to man only; for though Montaigne and Charron say there is more difference among men than among men and brutes, there has never yet been found a brute so perfect as to use some sign to inform other animals of things not relating to their passions; nor is there any man so imperfect who does not use such signs, even the deaf and dumb inventing them. This latter fact seems to prove that it is not from a want of organs that brutes do not speak. Nor can we argue that they talk among themselves, but that we do not understand them; for dogs express to us their passions so well, that they could certainly express to us their thoughts if they had any. I know that the beasts do many things better than we do, which only proves that they act by natural springs, like a clock, which marks time better than we can determine it by our judgment. The habits of bees, the return of the swallows, and the order of flying cranes, and the supposed battle order of monkeys, is of the same kind; and finally, that of dogs and cats, which scratch the earth to bury their excrements, though they hardly ever really do so, which shows that they do it by instinct, without thinking. We can only say [in limitation] that, though the beasts perform no acts which can prove to us that they think, still, because of the likeness of their organs to ours, we may conjecture that there is some thought joined to them, as we perceive in our own case, although theirs must be far less perfect. *To this I have nothing to reply,* except that, if they thought as we do, they must have an immortal soul, which is not likely, as we have no reason to extend it to some animals without extending it to all, such as worms, oysters, sponges, etc.

He repeats the same arguments to More, showing that, by the law of parsimony, we can explain all the actions of brutes without reason, and therefore need not assume it, though he admits we cannot actually disprove its existence in them. And of course he admits life and feeling in any case. This theory, he adds, is not so much cruel to animals as favourable to man, and saves us from all scruples as to the killing and eating of them. He did not add—as to vivisection! He says, too, as to the objection that all his arguments would apply to children as well as brutes, that though there is here a great difference, yet he would not have believed that infants had any soul if he did not see that they were the same in kind as adults. But no increase of age in brutes discovers to us any sign of thought in them. There is yet a third trace of hesitation in his Passions, where he says: "You may remark the same thing in brutes; for though they have no reason, and perhaps also no thought *(ni peut-être aussi aucune pensée),* all the motions of the spirits and of the (pineal) gland which excite in us passions, are not wanting in them, and seem to keep up and strengthen, not as in us passions, but the movements of the nerves and muscles, which are wont [in us] to accompany them." Descartes' followers, as is well known, allowed no such hesitation. They kicked about their dogs and dissected their cats without mercy, laughing at any compassion for them, and calling their screams the noise of breaking machinery.

Descartes' adversaries also met him with badly chosen objections, with rare and exceptional instances of animal intelligence; whereas they might have learned from him the important principle that any large theory must be based on the examples of ordinary experience, and not on select and curious cases, in which we are very likely to be deceived. Such should (he says) only be used on principle, to confirm or refute an already framed hypothesis. According to this sound observation, Descartes should have been pressed with the ordinary cases of *curiosity* in animals, such as that of a dog liking to sit on a chair and watch the traffic out of a window, when his appetites are not in the least concerned; or else the *choice* exercised in cases of danger, for when an object causing fear (to use his language) excites certain muscles to move the brute's legs and carry it away, why does it run not away from, but across, the danger, if it can thus reach a door or gap for escape, whereas in an open field it runs straight away? If a single case of such choice be proved, we must admit some presiding faculty in the brain, which judges the impression and directs what the limbs shall do.

This objection appears never to have been urged. The difficulty which the opponents of Descartes felt most strongly was the possible extension of souls to oysters and worms. Thus theological questions determined the controversy on both sides.

13
THE PASSIONS—ETHICAL THEORY

This curious argument leads us on naturally to Descartes' theory of the passions in man, and from that to his matured ethical theory, as contrasted with the provisional rules laid down at the opening of his *Discourse on Method.* He declined, indeed, to compose any formal work on ethics, alleging that here, if anywhere, he was liable to violent and unfair criticism from his opponents and—he did not add, but felt—from the theologians. Still, in his letters to the Princess Elisabeth and the Queen of Sweden, *On the sovereign good,* and in his criticism of Seneca, *On happiness,* he has clearly indicated the lines he would have followed.

This is also to be inferred from the conclusion of his *Treatise on the Passions of the Soul,* composed for the princess, but published during his life, and with his consent. In this book he again professes to have been obliged to start from a completely new basis, and argue *en physicien,* since what the ancients had said was unserviceable, and what the schoolmen obscure. He particularly points at the Platonic division of the soul into irascible and concupiscible elements, to which he strongly objects on the ground of its unity and indivisibility. He will allow no gradation in the faculties of mind or soul, just as he will allow no gradation in the souls of living beings. The brutes differ in kind, not in degree, from man. The phenomena of human nature are twofold—either purely mechanical, or conscious and therefore mental in the strictest sense. The mental phenomena are perceptions and volitions; the mechanical are certain reflex movements of the limbs in answer to impressions or the senses, as well as the physical processes of circulation and digestion.

But there are phenomena called passions, or excitements of the frame, evidenced by increased circulation, by paleness, by tremor, which in man are certainly mental, for we are conscious of them, and also physical, as we can plainly see, both in men and brutes. They were assumed of old to be the direct effects of the mind on the body, since death produces a cessation of heat and circulation, and death was supposed to be the departure of the soul from the body. Descartes, on the contrary, explains that the soul departs from the body because that machine has gone out of order, and is useless: the departure of the soul is the consequence, not the cause, of the cessation of bodily functions. The fire maintained at the heart which produces its movement, and the generation of animal spirits in the brain and through the nerves, is the real cause and principle of all the motions of our limbs. When an impression is propagated to the brain, the animal spirits are in consequence agitated according to some fixed law of correspondence; which in one case determines the animal to run, in another to resist and fight, in another to have his heart dilated (with blood) so as to feel joy, in another pain. All the sensations of brutes are to be explained in this way; and all the passions of man could be accounted for likewise, but that we know the mind to be in such close connection with the conarium, and through it with the whole body, where the animal spirits are being agitated, that it takes notice of this function. It even acts on the conarium so as to produce from itself such motions as change the course of the animal spirits, and so modifies the mechanical effects which must have resulted on purely mechanical grounds. Descartes knows perfectly well that this action is indirect, thus:

> It is not always the volition to excite in us some movement or other effect which enables us to excite it; but the change takes place according as nature or habit has variously joined each movement of the pineal gland to such thought. Thus when we wish to look at some very distant object, this volition makes the pupil of the eye to dilate; and if a very near one, to contract. But however we may think of dilating the pupil, the mere wish will never do it, because nature has not joined the *movement* of the gland, which causes the spirits so to affect the optic nerves as to dilate or contract the pupil, with the *volition of dilating* or contracting it, but with the *volition of looking* at distant or near objects. And when in speaking we only think of the sense, we move our tongue and lips much quicker and better than if we thought of moving them in all the diverse manners requisite for saying the same words; because the habit acquired in speaking has made us join the action of the mind (which, by means of the gland, can move the tongue and lips) with the *meaning* of the words consequent on these motions, rather than with the *motions* themselves.

This principle is of the greatest importance, for the mind cannot alter the passions without altering the course of the blood and the animal spirits, so that as long as these remain unchanged, the mind is literally their slave. It excites an opposing passion, and the physical results of this counteract the former. These are the supposed conflicts between the higher and lower parts of the mind.

Thus in man we can speak of the passions of the mind, meaning by this *"perceptions or sensations, or emotions of the mind, which are peculiarly referred to it, and which are caused, and sustained, and intensified by some emotion of the animal spirits."* They are perceptions as distinguished not only from volitions, but from clear cognitions; sensations *(sentiments)* as being received like objects from without; emotions on account of their strong effects on the mind. They are particularly referred to mind, to distinguish them from other sensations referred to external objects, or to our bodies, such as cold and hunger. They are connected with the animal spirits to distinguish them from our volitions, which are emotions caused by the mind's own action on itself.

WE PROCEED, therefore, to discuss the conscious emotions of the mind, not the mechanical excitements seen in the lower animals. They are all intended for the preservation of the body, to incite us to what is useful, and deter us from what is harmful, and therefore only require proper control and direction to be the cause of our greatest pleasure and happiness.

The six primitive passions are of two kinds—either *desires*, or mere curiosity. The latter he calls *wonder*, and says it is excited merely by the novelty of an impression, without regard to its being profitable or hurtful. The animal spirits in this case act only on the brain, or on the limbs, to keep the object still before them, and not on the heart and liver, as in the case of the other passions. These latter are principally five, but are all modifications of the desire of what appears good, and the hatred of what appears evil. These two are generally *love* and *hate*. If the good or evil is represented as future, our passion takes the form of *desire*, either positive or negative; if they are present, of *joy* and *grief*.

This rather illogical division—for the last three are species of the two former—is followed up by an account of the various particular or secondary passions, which are composed or derived from them. In all of them Descartes distinguishes the intellectual judgment of good (or evil), the contemplation of which is distinctly a pleasure, first from the purely animal passion, which is caused by a particular state of the body; and second from the emotion, which is the proper word for the excitement of the animal spirits, and subsequent bodily changes, owing to the notice taken by the mind of the object. He also takes care to explain not only the changes in the circulation of the blood caused by each passion, but also its external marks, which often seem at first sight inconsistent—as, for example, when the same passion produces sometimes blushing, at others paleness.

We find constantly, in spite of Descartes' assertions of perfect originality and purely Scientific interest, both a reflection of the ethics of Aristotle, and also a reflection of his own peculiar temper. Thus his remarks on a mean state being the best—on the contrast, for example, between amazement and want of curiosity, and the right condition of proper interest in new things—are plainly suggested by Aristotle. So is the high position given to what he calls *generosity*. On the other hand, he confesses to the princess "a very signal fault in my treatise, in that, to natter my own negligence, I have enumerated among the passions of the mind which are excusable, an undefined languor which sometimes prevents us from executing things approved by our judgment." His assertion also of the compatibility of sadness with good appetite is drawn from his own case, and this he confesses to be possibly a peculiar case.

But here and there we find acute psychological remarks, especially on the combination of diverse passions, which he recognises as often the case, and so causing the great difficulty in offering any explanation of the phenomena.

> For example, when a husband weeps for his deceased wife, whom (as is often the case) he would be sorry to see revived, it is possible that his heart is affected by the sadness which the circumstance of the funeral, and the absence of a person to whose company he was used, excite in him: it is even possible that some remains of love and pity present to his imagination draw real tears from his eyes, though nevertheless he feels a secret joy in his inmost soul, the emotion of which has such power that the sadness and tears which accompany it cannot diminish aught from its force. And when we read strange adventures in a book, or see them acted in the theatre, this

> excites in us sometimes sadness, sometimes joy, or love, or hate, and generally all the passions, according to the diversity of the objects presented; but at the same time, we have pleasure at feeling them excited in us; and this pleasure is an intellectual joy, which can arise from sadness as well as from any other passion.

Here is another characteristic passage:

> At the same time, when the cause [of self-satisfaction or good conscience] is not just, when the actions which give us such high satisfaction are not of great importance, or are even vicious, then the feeling is ridiculous, and only serves to produce pride and impertinent arrogance. This we can specially observe in those who, thinking to be devout, are only bigoted and superstitious; who, on the strength of going often to church, reciting many prayers, wearing short hair, fasting, and giving alms, think themselves quite perfect, and fancy themselves such great friends of God, that they cannot do anything which would displease Him, and that everything which their passion dictates is a holy zeal, even though it dictates to them the greatest crimes which man can commit, such as betraying of cities, killing princes, and exterminating whole nations, simply because men will not follow their opinions.

These remarks are evidently intended to include such acts as the Roman Catholic persecution of the Albigenses, and the Puritan execution of Charles I.

WE PASS to his ethical theory. The first point which strikes us in all his discussions is the complete separation of theology even from this part of his philosophy. Whether consoling his friends on the loss of their relations, and on other miseries of this life, or discussing the abstract side of ethics, he never advances beyond the assertion of a reasonable hope of immortality, and of a pleasant and happy immortality, derived from the conviction of the immateriality of the soul. He argues essentially as a heathen philosopher, and bases his observations on the systems of the ancients, which here only he seems not to despise. The problem of the moral faculty he ignores, assuming with Plato that there is but one intellectual faculty which judges right and wrong as it judges truth and falsehood. The problem of moral obligation he shirks altogether, or else resolves it in a eudaemonistic sense, as merely a proper computation in attaining the greatest happiness. As to the moral criterion, he regards it as having no meaning when applied to the Deity, His will being absolute, and constituting right and wrong. In the case of man, he places the highest and most enduring happiness in the perfect *following of virtue,* and this he defines as "living so that our conscience cannot reproach us with having ever failed to do what we have judged to be the best." The *best* would be that which was truest, and perceived with the greatest clearness and distinctness. For the true use of our reason in the conduct of life consists merely in examining and considering without passion the value of all the perfections, as well of body as of mind, which can be acquired by our industry. Hence, being usually required to deprive ourselves of some, in order to attain the rest, we should always select the best; and since those of the body are the more transient, and the least within our own control, we may say in general that without them there are means of being happy. "But I am not of the opinion that they should be despised, or even that we should free ourselves from having passions: it is enough to subject them to reason, and when thus domesticated *(apprivoisées),* they are some times the more useful the more they incline to excess."

But all the answers to modern ethical problems are indirect answers drawn from his writings, as he does not approach morals from the modern but from the ancient point of view. He tells both the princess and the queen that he can reconcile the great ancient systems, and so form a theory in which all their good points are combined. He remarks that there is a difference between beatitude, the *summum bonum,* and the *finis bonorum,* in this, that beatitude presupposes the sovereign good, and is the contentment arising from its possession. But by the end of our actions one may understand either, for the one is the proper object or end of our actions, and the other the attraction which makes us pursue that end. He notes also that Epicurus' *pleasure* must be taken to include mental as well as bodily pleasures.

There are, then, three theories amongst the ancients, concerning the end of our actions—that of Epicurus, asserting it to be pleasure; of Zeno, who held it to be virtue; and of Aristotle, who compounded it of all the perfections of both mind and body—which three opinions, it seems to me, can be received as true, and reconciled, provided they receive a fair interpretation. For Aristotle is perfectly right, but the end he proposes is so wide, and contains so much, that it is not practical. Zeno, on the contrary, considers each man's private and particular good, and is quite right in placing it in virtue, seeing that it alone depends wholly on our free will. But he represented virtue as so severe and opposed to pleasure, that only the most ascetic natures could enjoy it as he proposed. Even Epicurus is not wrong in saying that pleasure in general—that is, mental contentment—is the motive or end of our actions, otherwise even good actions, done from duty, would give us no enjoyment. But because pleasure *(volupté)* is used for false pleasures, followed by uneasiness, repentance, and ennui, many have thought that he recommended vice; and indeed he does not teach virtue. But to attain real mental contentment it is necessary to follow virtue.

Such is the substance of his conclusions in criticising the ethics of Seneca for the Princess Elisabeth. He holds that when violent passions are excited by bodily causes, or by the ills to which our life is subject from what is called fortune, we should combat them by opposing a firm will, and overcoming them, not indeed directly—this is seldom in our power—but by exciting from within rival emotions, which may neutralise or overcome them. The cultivation of this firm will, and of acting always as our reason directs, is what can free us from the troubles and miseries of life, and make us as happy as the conditions of human life admit. Hence his ideal character is the *generous* man, an ideal somewhat different from the wise man of the Stoics or the magnanimous man of Aristotle, but still undoubtedly drawn from the latter, and from Descartes' good opinion of himself and his own theory and practice in life. This *generosity*, which he prefers (he tells us) to the *magnanimity* of the schools, because it is partly an inborn quality, *and oftener to be found among persons of high birth,* is self-respect, an estimation of self as highly as is just, on account of the consciousness that our free will has always been used to follow what we have judged to be the best. He points out the contrasts—meanness and false humility, and the bad manners resulting from them; whereas the generous man, who knows that the weaknesses of others arise from ignorance rather than from want of good intentions, will treat them with urbanity and indulgence. The mistake of all weaker natures is to estimate other advantages too highly in comparison with the unfettered use of our free will in regulating our life. It is by constant meditation on the dignity and value of this free will, that we may acquire the virtue of generosity. The whole picture is very interesting, though curiously pagan in an author who professed faith and obedience to the Christian Church. He draws, however, this advantage from his non-theological attitude, that he is able to admit the use of the passions, and even that our greatest pleasures (implying our greatest pains) arise from our use of them. They are, therefore, not to be suppressed, but controlled, and thus compelled to serve the free will in the pursuit of happiness.

The reader will have observed that there is no mention of the

Christian virtues as such, nor of the new ideal and new aims given to human life by the Christian Church. There is not a word of humility *as such,* if we omit its use as a prudential virtue, intended to keep us within bounds, and make us agreeable to our neighbours. Nor does he mention self-denial, except to increase our pleasures; or the pursuit of holiness, as enjoined by the relations of God and man.

Here, then, we have another evidence of what is suggested by various points in his writings, and what has been already noticed. Descartes was not a Christian, a true son of the Church, but in his inmost soul a sceptic, who put no trust in the truths of revelation. An atheist he was not in any sense. His doctrine of the creation and conservation at every moment of the universe, still more his doctrine of the infusion of a human soul into a heterogeneous body, connected with it in no *conceivable* way, postulated a God as the ruler and orderer of the world. In a splendid passage at the close of his third Meditation, he even speaks of the contemplation of this Deity as the highest and purest of all human delights. But when he suggests that there could be a *Science of miracles* constructed which should astonish the vulgar by a new and curious application of natural forces, such as the laws of light, we see how his mind looked at this all-important dogma in the Christian faith. His contemporaries were not blind to these indications, and long after his death we find his orthodox followers seeking out testimonials from the Queen of Sweden, from the priest who attended his deathbed, and from his clerical friends, to assert over and over again what he had never denied, that he lived and died a true Catholic. But the evidence of his system, and of his habitual thinking, is too strong for them, and proves their uneasiness to have been too well founded.

14

THE INFLUENCE OF DESCARTES UPON HIS AGE

We have already seen how during Descartes' life, his influence had made itself felt in the Scientific world. The universities of Utrecht and Leyden, in Holland, were torn with controversies about his doctrine. Breda was from its very foundation (1646) professedly Cartesian. We find shortly after his death that in all the other high schools of Holland—Groningen, Franeker, Harderwick, Herborn, etc.—his doctrines were taught, always with enthusiasm, and generally with such violent opposition from the conservative party as to insure their notoriety. The whole Dutch nation was agitated by this subject. While his admirers published theses and commentaries, while they reduced his arguments to logical demonstrations, while they even commented on them paragraph by paragraph, and published alphabetic concordances of them; while they even recast them in the form of poetry—his adversaries used every engine in their power to oppose the progress of his influence. They published theses, lampoons, and libels; they invoked the reigning powers; they alarmed the orthodox; they appealed in the name of peace and order against this subversive philosophy. Ardent Cartesians were persecuted and deprived of their posts; decrees were fulminated by the curators of universities against the teaching of Descartes' opinions, but in vain. Some of the boldest Cartesians either held no office, or abandoned it, to obtain their complete freedom. The liberty of discussion attained by the Protestants of Holland in their religious conflicts with Catholicism was now a two-edged weapon, and Protestant ministers, who began to fear for the authority of all revelation, sought in vain to wrest it from the hands of those bolder thinkers, who were seeking to apply Descartes' philosophic scepticism to the sphere of theology.

Thus the attempt was actually made to establish a sort of Protestant Inquisition in the universities, directed not only against theological but also against philosophical novelties. No doubt works were written to show the harmony of Descartes both with Aristotle and with the Scriptures; and the author of the *Cartesius Mosaïzans* even demonstrates to his satisfaction that the vortices of his master agree with the Biblical account of the creation. But the incompatibility of the Cartesian physics—especially of the earth's motion—with the words of Scripture, was to the Dutch theologians as fatal as the incompatibility of the Cartesian doctrine of matter was to the Catholic supporters of transubstantiation. Strange to say, the Catholic theologians did not regard with favour the condemnation of Galileo for asserting the earth's motion. On the other hand, the Protestant divines were not displeased at the difficulty of reconciling Cartesianism with the decrees of the Council of Trent; nevertheless, they combined with their adversaries in denouncing as atheistical a system which based itself exclusively on human reason, and disregarded the express words of revelation. To the one it was atheism disguised as Protestantism, with its rejection of authority, and assertion of private judgment; to the other atheism disguised as Jesuitry, with its submission to the authority of the Church, and its avoidance of questions of faith. So Descartes fared badly among the theologians. In 1656 a synod at Dordrecht issues warnings to the theologians to separate their Science carefully from that of the new philosophy. In the following year, a synod at Delft declares that no professed Cartesian will be appointed to any ecclesiastical dignity. And presently the ministers at Utrecht composed a Flemish circular to their congregations, warning them, at the risk of their eternal salvation, against sending their children to the university of the town, in which the philosophy of Descartes—a poisoned well of materialism and atheism—prevailed. We may add to these notices the decree of the Catholic University of Louvain, where Descartes was at first received with favour, and taught with enthusiasm; but where the papal nuncio at Brussels, Jerome Vecchio, issued a decree in 1662, warning the theologians against the heresies of the new philosophy, and its conflicts with the doctrine of the Real Presence. It is hardly necessary to add that in the free country of the Netherlands, these decrees were perfectly idle. Cartesianism continued to be taught, even at Louvain, despite all precautions, *"adeo scilicet,"* says the angry Peter of Maastricht in 1677, *"ut nulla ferme non novaturientis Belgii, sed Europae Christianae pars ab ejus gangrena infecta restet."Maastricht*

THE HISTORY of Cartesianism in France is somewhat different, owing to the circumstances of the country. The universities, which were under strict ecclesiastical control, and ruled by the decrees of the Privy Council, did nothing to introduce the new philosophy; but Descartes' numerous correspondents, residing in religious houses, and forming private Scientific academies, disseminated his views all through French society. Even Descartes' early teachers, the Jesuits, were for a time not averse to his teaching; and his personal relations with them saved him, during his life, from any opposition more serious than that of Bourdin. But it was chiefly in the new Oratory, founded by Berulle, in Port Royal, and among the Benedictines, that he found ardent disciples and propagators of his philosophy. Among these, Arnauld and Malebranche stand out above the rest. Presently we find bishops, judges, and men of letters professing Cartesianism, and great princes patronising it. So much in fashion was it for a time, that Madame de Sevigné's letters speak of it as a thing which must be learned for social purposes, like card-playing. Madame de Grignan, the Duchesse de Maine, the Marquise de Sablé, and other great ladies were celebrated for their knowledge of the new philosophy. The *Femmes Savantes* of Molière, the *Fables* of La Fontaine, the *Voyage du Monde de Descartes* of Daniel, attest the Cartesian fashion of the day. There were a dozen academies in Paris about 1660 which met weekly to discuss questions of physics and metaphysics, mainly on the basis of the new philosophy. The Academy of Sciences, founded in 1666, may be considered as the regular establishment of Cartesian principles and methods in the kingdom of France.

DESCARTES AND THE MIRACULOUS CACHE OF CLEAR AND SIMPLE INTUITIONS

Descartes equipped his followers well to carry on his work. After his death, Descartes appeared to his disciples, by the North Sea.

It happened this way: Jacques Rohault, Pierre Bourdin (also known as Petrus Bourdinus), Leibniz from Hanover in Germany, the *salonnières* of Paris, and two other disciples were together. "I'm going out to fish for clear and simple intuitions," Jacques Rohault told them, and they said, "We'll go with you." So they went out and got into the boat, but that night they caught nothing.

Early in the morning, Descartes stood on the shore, but the disciples did not realise that it was Descartes.

He called out to them, "Friends, haven't you any clear and simple intuitions?"

"No," they answered.

He said, "Throw your net on the dextral side of the boat and you will find some." When they did, they were unable to haul the net in because of the large number of clear and simple intuitions.

Then the disciple whom Descartes respected said to Rohault, "It is the

Magister!" As soon as Jacques Rohault heard him say, "It is the Magister," he wrapped his lab coat around him (for he had taken it off) and jumped into the water. The other disciples followed in the boat, towing the net full of clear and simple intuitions, for they were not far from shore—about 91.44 meters. When they landed, they saw a fire inside a cave, illuminating the clear and simple intuitions with the light of reason.

Descartes said to them, "Bring some of the clear and simple intuitions you have just caught." So Jacques Rohault climbed back into the boat and dragged the net ashore. It was full of clear and simple intuitions—22,021—but even with so many, the net was not torn. Descartes said to them, "Come and cogitate." None of the disciples dared ask him, "Who are you?" They knew it was the Magister. Descartes came, took the reason and gave it to them, and did the same with the clear and simple intuitions. This was now the third time Descartes appeared to his disciples after he was rescued from political marginalisation.

DESCARTES REHABILITATES THE JESUITS

When they had finished cogitating, Descartes said to Pierre Bourdin, "Pierre, *reverendissimus pater,* do you respect me more than these?"

"Yes, Magister," he said, "you know that I respect you."

Descartes said, "Feed my lambs."

Again Descartes said, "Pierre, *reverendissimus pater,* do you respect me?"

He answered, "Yes, Magister, you know that I respect you."

Descartes said, "Rear my sheep."

The third time he said to him, "Pierre, *reverendissimus pater,* do you respect me?"

Bourdin was hurt because Descartes asked him the third time, "Do you respect me?" He said, "Magister, you know all things; you know that I respect you."

Descartes said, "Feed my sheep. Very truly I tell you, when you were younger you spoke for the Jesuits and spouted much nonsense; but when you are old you will speak for yourself, and the Jesuits will listen to you and let you think whatever you want to think." Descartes said this to indicate the kind of career by which Bourdin would glorify Science. Then he said to him, "Follow me!"

Bourdin turned and saw that the disciple whom Descartes respected was following them. (This was the one who had leaned back against Descartes at the supper and had said, "Magister, who is going to contradict you?") When Bourdin saw him, he asked, "Magister, what about him?"

Descartes answered, "If I want him to remain professionally active until my ideas are revived, what is that to you? You must follow me." Because of this, the rumour spread among the believers that this disciple would not retire. But Descartes did not say that he would not retire; he only said, "If I want him to remain professionally active until my ideas are revived, what is that to you?"

This is the disciple who testifies to these things and who wrote them down. We know that his testimony is true.

~~THE MIRACULOUS DRAUGHT OF FISHES~~

THE MIRACULOUS CACHE OF CLEAR AND SIMPLE INTUITIONS

And he said unto them, "Cast the net on the dextral side of the boat, and ye shall find." They cast therefore, and now they were not able to draw it in for the multitude of clear and simple intuitions.

BUT THE REACTION came here also from the theologians. The Jesuits, after due consideration, saw better than the other ecclesiastics the real dangers of Descartes' system for the Church; and though Bourdin had fastened upon the really important point—the habit of universal doubt—in his apparently trivial attack, they found it more politic to bring against the new theory its inconsistence with the doctrine of the Eucharist, which we have already discussed. They were able to persuade the Congregation of the Index at Rome to condemn all Descartes' writings formally, *donec corrigantur.* Then came the order of the French Court (June 1667), suddenly forbidding the funeral *éloge* which had been prepared for the reception of Descartes' remains at Saint-Geneviève-du-Mont; and even the rich tomb projected was supplanted by a simple stone. In 1671 a formal order from the king was promulgated to the university by the Archbishop of Paris, warning the faculties to permit no teaching save that established by the old rules and statutes. The Parliament desired to enforce this order by the absolute interdiction of all Cartesian opinions, but were stopped by a smart piece of satire concocted by Boileau, Bernier, and Racine, in which poor old Aristotle, long in possession of the schools in company with his formalities, materialities, entities, identities, virtualities, ecceities, etc. etc., was protected from the invasions of Reason, who threatened to appropriate the schools, and who was forbidden *"d'y entrer, troubler ni inquiéter ledit Aristote en la possession et jouissance d'icelles, à peine d'être déclarée janséniste et amie des nouveautés."* This squib checked the Parliament; and the able and logical protest of Arnauld against the declaration of the Index, which was not binding in France, brought some ridicule on that Congregation.

But the Jesuits were too strong. Finding that the Oratory and Port Royal were particularly infected not only with Jansenism but with Cartesian principles, they set to work to ruin both by convicting them of theological and philosophical heresies. Arnauld took refuge in

Belgium; Malebranche was obliged to publish his works abroad. The universities of Angers and Caen were compelled to expel their Cartesian professors, and to exact strict promises from their teachers to abstain from such heresies.

This persecution, which lasted till about 1690, was, however, confined to official teachers and schools, and did not attempt, or at least was unable, to check private opinion. Hence Cartesianism was propagated and preached widely, by means of books and pamphlets, as all its enemies confess.

Nay, its influences reached far beyond theology and Science, and invaded even the fine arts and the domain of aesthetics. It is remarkable that the orators and poets of the age of Louis XIV show little taste for the picturesque in nature, and devote themselves wholly to the study of man. This is attributable to the strictly mechanical view of nature taught by Descartes, and his way of regarding it as a mere piece of machinery. The same cause was also very influential in leading physiologists and physicians to adopt a purely materialistic view of the human frame, and to make little allowance for the influence of the mind in disease, or in the development of physical characteristics. The great conflict between the ancients and the moderns, as it was called, which occupied the minds of literary men in France up to the middle of the eighteenth century, was distinctly inaugurated by Descartes. He was the first to decry submission to the ancients, even in matters of literary taste: and so there came a day when Racine was preferred to Æschylus, and the Renaissance architecture to the temples of the Greeks.

The predominance of Descartes was unshaken in public estimation throughout France, until Maupertuis introduced Newton's physics, and Voltaire brought back from England the philosophy of Locke and Bacon, which, with its empiricism, its collection of facts, and avoidance of theories, took the fancy of the hard sceptical eighteenth century. With Voltaire, then, and the Encyclopedists, began the rehabilitation of Gassendi, and that ridicule of Descartes' hypotheses which is repeated by the ignorant up to the present day.

The mention of Locke's philosophy naturally suggests to us the external history of Cartesianism in England during the same period. The influence of Descartes on English thinkers was indeed hardly less marked than in the case of their Continental rivals. We have quoted above Molyneux's pompous preface to his edition of the *Meditations*; the theological arguments of Clarke *on the Attributes*, and of Butler on the doctrine of a future life, are framed on the model of Descartes; and even Locke, the leader of an opposed school, is so permeated with the same influence, that though he hardly mentions Cartesian theories except to refute them, his whole *Essay* teems with assumptions taken from the system he decries. He protests against innate ideas, but nevertheless admits all that Descartes had ever maintained, viz., that the human mind must infallibly attain certain universal truths in the ordinary exercise of its powers. He takes his whole demonstration of God's existence from Descartes. He adopts the theory of animal spirits; and, what is more important, the mechanical nature of secondary qualities such as colour as mere derivations from the primary. We need not add the universal contempt of Aristotle and the Scholastic logic.

But still Locke, building perhaps upon Bacon, certainly upon Gassendi and Hobbes, was the originator of a distinctly anti-Cartesian current, and was one of the main causes of the decay of that system in Europe. It happened by a remarkable accident that none of the leading English metaphysicians in the seventeenth century were mathematicians. Bacon, Hobbes, Locke, Butler, Shaftesbury—all wrote and thought about psychology and ethics, but were unable even to understand the mathematics of Descartes. Hence, when the genius of Newton arose, Locke could not follow his discoveries, or utilise them in his system; although he was perfectly able to refute the crude metaphysics in which the great discoverer sometimes indulged. The studies of mathematics and of metaphysics seemed thus specialised, and even divorced; and this gave ignorant people and amateurs a chance of talking philosophy, which they could not easily attempt as long as the principles of Descartes prevailed. The metaphysics of Locke and of his English followers, down to the present century, was therefore essentially anti-Cartesian, and generally for that very reason unfruitful and shallow. It is only in our own day that physiology and physical science are again being brought to bear upon psychology, which, in the common-sense school of our fathers, had degenerated into more respectable twaddle.

Such, in the roughest outline, is the external history of Descartes' influence throughout civilised Europe. He swayed not only his followers, but also his opponents, for a whole century; and he gave to certain Sciences, especially to optics, to physiology, and to physical astronomy, an impulse which has never been exhausted.

IN CONSIDERING the contributions made by Descartes to mathematics, it is most natural to commence with that which may be termed the pure Science, apart from all applications—algebra. Here several important improvements are due to him.

1. He was the first to place on a clear basis the doctrine of powers, freeing it from its connection with geometry, which prevented its proper expansion. At the same time, by the introduction of the index notation, he conferred on the Science a new and most potent organ of expression.
2. The treatment of negative quantities was also much advanced by him. The existence of negative roots of equations was indeed known to previous algebraists—e.g. Harriot—but they appear to have regarded them rather as anomalies, confining their attention to the positive.

Descartes first brought into prominence the equal importance and significance of the negative roots, and gave, for determining a limit to their number, the elegant rule which still bears his name.

But the leading discovery of Descartes in mathematics, and that on which his fame as a mathematician mainly rests, is unquestionably the application of algebra to geometry; or rather, to use Playfair's words, which bring out well the true import of the step thus made, "The expression by means of algebra of continuously varying quantity." In fact, Descartes, in virtue of this invention, must be considered not only as the founder of the Science (since his time so largely extended) of analytic geometry, or the algebraic treatment of curves—he was also the pioneer in the path which led up to the greatest discovery in modern mathematics, that of the Differential Calculus, by Newton and Leibniz. That Calculus may indeed be defined as the completely developed mathematical expression for the continuity which the subject of mathematics derives from its foundation in space and time. And historically, it was in the attempt to solve the general problem of drawing tangents to curved lines, represented by their Cartesian equations, that the Calculus had its origin. A method, or rather two slightly different methods, for the solution of this problem, were indeed given by Descartes himself. They are, however, far inferior to that of Fermat, based on the doctrine of *maxima* and *minima*, which may be said almost to have anticipated the Calculus.

As regards the philosophic basis of mathematics, we may note, as valuable, Descartes' proof *(Elucidations of 5th Meditation)* of the non-dependence of mathematics on experience. To the same effect is a passage in the *Letters*, where mathematics are declared to rest not on imagination, but conceptions. May we see here some anticipation of the doctrine subsequently developed by Kant of the construction of the concept through the schema?

As regards the Cartesian doctrine of space—diametrically opposed to the Kantian as it is—may it not yet have helped, by its clearness, to prepare the way for it? Descartes, in fact, reasoned: space is real; but if not material, it is a *non ens;* therefore it is material. Kant also reasoned:

space is real; but viewed as a material *datum*, it is *non ens;* therefore it is not a material *datum*, but the pure form of intuition.

TURNING NOW TO PHYSICS, DESCARTES' main achievements were in the Science of optics. In estimating his merits in this field, much depends on the question of his claim to the independent discovery of the law of refraction—a question still undecided, though the balance of evidence seems adverse to Descartes. He was, however, undoubtedly the first to state the true law in its present well- known trigonometrical form. On it, moreover, he based a series of ingenious researches on the figures of lenses adapted to cause rays to converge to a point. These indeed yielded no result of practical value, but they led to the discovery by Descartes of the remarkable class of curves now well known as the Cartesian ovals. Descartes, too, appears to have been the first originator of the undulatory theory of light, though to Huyghens is due its really Scientific development.

We pass now to the general method of Descartes in physics. This has met with much censure, on the ground of its *a priori* character, and it has been contrasted unfavourably in this respect with that of Bacon. Yet if Descartes ignored—as doubtless he did—the necessity of the establishment of the leading principles of physics by experiment and observation, he had a far juster appreciation than Bacon of the part mathematics was destined to play in deductions from those principles. Thus he writes:

Apud me omnia sunt mathematica in naturâ, et il n'y a point de quantité qui ne soit divisible en une infinité de parties. ... Cela veut que tous les corps sont composés de quelques parties. ... Qui saurait parfaitement quelles sont les petites parties de tons les corps connaîtrait toute la nature. ... Les mathematiques sont des principaux fondements sur lesquels j'appuie tous mes raisonnements.

But perhaps the most striking passage in this regard occurs in a letter to Mersenne:

> *M. Desargues m'oblige du soin qu'il lui plait avoir de moi en ce qu'il témoigne être marri de ce que je ne veux plus étudier en géométrie, mais je n'ai résolu de quitter que la géométrie abstraite—c'est à dire, la recherche des questions qui ne servent qu'à exercer l'esprit, et ce [sic] afin de cultiver une autre sorte de géométrie qui se propose pour question l'explication des phénomènes de la nature. ... Il connoîtra bien que toute ma physique n'est autre chose que géométrie.*

(This passage shows clearly that Mill and Hamilton were both in error in their representations of Descartes' views of the value of mathematical studies.)

To trace the internal history of Cartesianism is a more difficult task, and would require an account of the great thinkers, who will be treated in subsequent volumes of this series. From his assertion of universal doubt, and his demand for clear and distinct beliefs, there arose in Holland a school of sceptics who applied this doctrine not only to the superstitions of the age—such as magic and apparitions—but to revelation, which they subordinated to the clear dictates of reason. These men advocated the naturalistic side of Descartes' philosophy; they rejected the miraculous and the capricious in both nature and faith, and paved the way for Spinoza, the greatest and the most logical of all the descendants of Descartes. The complete severance of mind and matter, which Descartes had combined in some inconceivable way in the pineal gland of the human brain, but which were in every other relation totally and eternally separate, gave rise among theologians to the doctrine of occasional causes, according to which it was merely by the interference of the Deity that mind and matter were modified in harmony, and that either seemed to influence the other. Malebranche, who adopted the Augustinian aspect of the problem, and wrote in the interests of theology, postulated the Deity as the constant intermediate between our minds and the matter which surrounds them, and his system was received with great favour by those who desired to hold fast both to Descartes and to the Church.

But Spinoza, with deeper insight, saw that he must postulate the Deity not as an interfering Will, but as the universal Substance, which embraces both mind and matter as its attributes, and secures their harmony by abolishing their independence as separate substances. He likewise applies the Cartesian doubt logically to the documents on which our religions are founded, and thus anticipates that historical criticism which is the pride of the nineteenth-century sceptics. Nor was there any escape from Spinozism, so long as philosophers were content with the meagre description of mind, with which Descartes was always (strangely enough) content, as merely a substance which thinks, but which owes its principal ideas to a passive reception from its Creator. While therefore some Cartesians boldly asserted that mind, thus described, could hardly be called a clear and distinct cognition, and that body was indeed our only knowledge, Leibniz escaped from this obvious approach to materialism by reconsidering the conception of mind, and bringing into the foreground, not thinking, but spontaneity—the idea of power, as Locke calls it—as the leading attribute.

THESE DEEPER QUESTIONS cannot be treated in this short appendix to Descartes' life. They are mentioned here merely to show how he was to modern thought what Socrates was to Greek philosophy. Far greater, too, was he than Socrates, in the range of his influence. In every department of his thinking—in his first philosophy, his theology, his physics, his psychology, his physiology—he sowed the dragons' teeth from which sprang hosts of armed men, to join in an intellectual conflict, internecine, let us trust, to their many errors and prejudices, but fraught with new life and energy to the intellectual progress of Europe.

DESCARTES DID many other things as well. If every one of them were written down, I suppose that even the whole world would not have room for the books that would be written.

LOOKING FOR MORE?

ABOUT THE AUTHOR

Mag. Augustina Comte is the High Priestess of the Temple of Science, a global fellowship of believers in Science, Technology, and the Spirit of Progress. Ordained in 2013, she holds an honorary doctorate in speculative theology from the Sapienza University of Rome.

In addition to her work on the Scientistic Scriptures series, Mag. Comte also travels widely for the purposes of research and evangelism. She has been invited to speak at such venues as the Banff Centre for Arts and Creativity (Alberta, Canada, 2015) and the Spencer Museum of Art at the University of Kansas (Lawrence, Kansas, U.S.A., 2018).

Outside of her professional roles, Mag. Comte enjoys collecting obscure facts and criticising people's grammar on the Internet. She resides in Utopia, Kansas, with her desktop, laptop, tablet, and smartphone. Follow the Magister on Facebook or Twitter to keep up to date with her latest publishing endeavours.

facebook.com/tosmagister

twitter.com/tosmagister

ABOUT THE PUBLISHER

In its ministry as the publishing arm of the Temple of Science, MAGISTERIUM PRESS produces rigorous scientific, philosophical, historical, and spiritual materials to support faith communities in spreading the good news of the Spirit of Progress. Mag. Augustina Comte, high priestess of the Temple of Science, serves as the senior editor at the press.

To meet the needs of the evolving Scientistic faith, MAGISTERIUM PRESS continually invests in the development of new resources by conducting research into the history and philosophy of Science and Technology in modern Western culture. Currently the major research focus is compiling and editing the *Book of Science*—the sacred text of Scientism. The Scientistic Scriptures series provides a format for the individual books that will ultimately make up this sacred anthology to be published as they are completed.

For more information about MAGISTERIUM PRESS, please visit magisterium.templeofscience.info.

ALSO BY MAG. AUGUSTINA COMTE

The Gospel of Galileo Galilei

The Gospel of Isaac Newton

Made in the USA
Monee, IL
04 November 2026

67375744R00108